Bromeliads for Home and Garden

UNIVERSITY PRESS OF FLORIDA

Florida A&M University, Tallahassee
Florida Atlantic University, Boca Raton
Florida Gulf Coast University, Ft. Myers
Florida International University, Miami
Florida State University, Tallahassee
New College of Florida, Sarasota
University of Central Florida, Orlando
University of Florida, Gainesville
University of North Florida, Jacksonville
University of South Florida, Tampa
University of West Florida, Pensacola

University Press of Florida

Gainesville Tallahassee Tampa Boca Raton Pensacola Orlando Miami Jacksonville Ft. Myers Sarasota

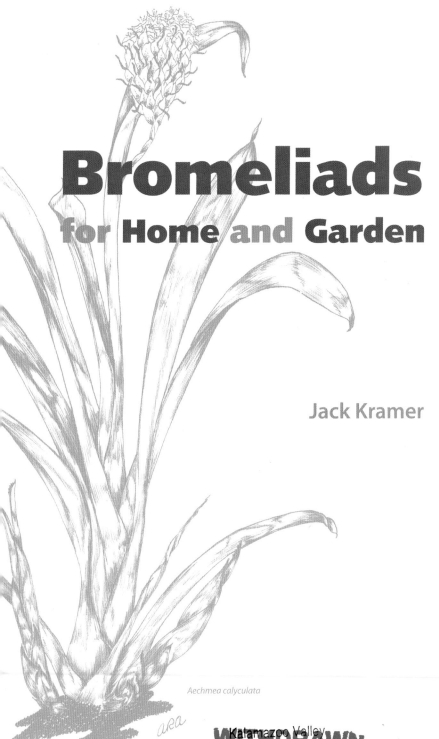

Bromeliads
for Home and Garden

Jack Kramer

Aechmea calyculata

Copyright 2011 by Jack Kramer
All line drawings by Andrew Roy Addkison
Printed in China on acid-free paper.

16 15 14 13 12 11   6 5 4 3 2 1

Library of Congress Cataloging-in-Publication Data
Kramer, Jack, 1927–
Bromeliads for home and garden / Jack Kramer.
p. cm.
Includes bibliographical references and index.
ISBN 978-0-8130-3544-4 (alk. paper)
1. Bromeliaceae—Florida. I. Title.
SB413.B7K694 2011
635.9'3485—dc22   2010032260

The University Press of Florida is the scholarly publishing agency for the State University System of Florida, comprising Florida A&M University, Florida Atlantic University, Florida Gulf Coast University, Florida International University, Florida State University, New College of Florida, University of Central Florida, University of Florida, University of North Florida, University of South Florida, and University of West Florida.

University Press of Florida
15 Northwest 15th Street
Gainesville, FL 32611-2079
http://www.upf.com

To Robert Curtis Alonzo

*Tillandsia
cyanea*

Vriesea hieroglyphica

Contents

Pitcairnia andreana

Author's Note

I started working with Bromeliads in late 1960, and in 1967 I did a small book on them as house plants, finding them a great companion for my orchid collection in my plant rooms in Illinois, California, and Florida. Bromeliads and orchids grow in nature under similar conditions. When I moved to Florida in 1985, I became intrigued with the wealth of Bromeliads, which were gaining popularity, and I decided to do a current book on the subject. I had dozens of Bromeliads growing with orchids in my landscape and plant room.

I had written 130 books on gardening subjects throughout the years, and now I felt a comprehensive book on Bromeliads was in order. The plants had gained much popularity since the 1980s, and I had learned a great deal of new information on them. This is my attempt to share my knowledge on these treasures of nature.

Orthophytum fosterianum

Preface

Bromeliads—Colorful Plants

Have you always yearned for brilliantly colored plants to add to the garden of your home or landscape? If so, Bromeliads are what you've been seeking. Bromeliads, generally characterized by stiff, vividly colored leaves often coupled with heads of bright flower bracts, provide the home gardener the double appeal of colorful flowers and berries, and ornamental foliage that last weeks.

Bromeliads exhibit an unbelievable variation in size (from miniatures to giants) and in color; leaves are found in every shade of green or yellow, brown, and red. Flower bracts range in color from red, yellow, blue, and green to pink, orange, lavender, and white. Bromeliads offer the possibility of year-round brilliance indoors or out.

I have grown some 200 Bromeliad plants under average conditions of apartment rooms and sun porches. I find they are exceptionally easy to care for and adapt faster and generally better than any other plant group.

To aid the gardener in beginning culture of Bromeliads, this book discusses which plants are rapid bloomers or need time for growth; which are suitable for confined spaces or need roomier settings; and which need direct light or can be successfully placed at normally barren northern exposures. (See the Quick Reference Guide at the end of this book.)

Such important facts as temperature, humidity, potting, watering, fertilizing, pests, and diseases are given coverage. Also included is a list of suppliers, a glossary, and a bibliography.

Taxonomists change Bromeliad names as research continues, so some of the names in this book may differ by the time this book goes to press.

Some of the plants referred to in this book or appearing in nature might be poisonous. Any person, particularly a novice gardener, should exercise extreme care in handling plants. Check with local authorities should you have any question about bringing a plant into your yard or home. The publisher and author accept no responsibility for damage or injury resulting from the use or ingestion of, or contact with, any plant discussed in this book.

Some of the material in this book has been used in my book *Bromeliads* (Harper & Row, 1981), now out of print.

1

Introducing Bromeliads

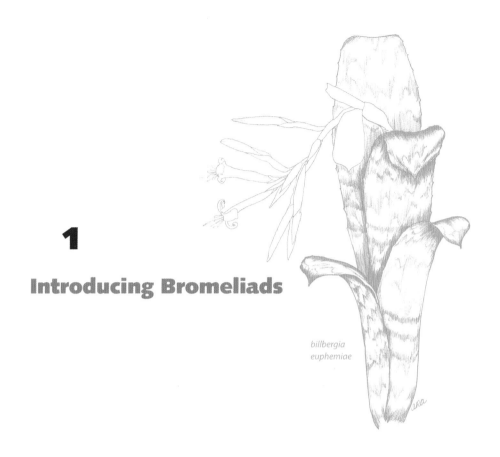

billbergia
euphemiae

About 3,500 species of Bromeliads are known, along with numerous hybrids—and more are introduced into cultivation yearly. But perhaps the best-known Bromeliad is the pineapple, *Ananas comosus*. Bromeliads are easy to cultivate and most of the plants have water reservoirs, which is handy if you tend to forget to water plants. Bromeliads will even survive for many months mounted on a tree without any water—such is their will to survive.

Most hobbyists find the classification of the Bromeliad family complicated, but it need not be. Basically *Bromeliaceae* is divided into three subfamilies: Pitcairnioideae, Bromelioideae, and Tillandsioideae.

Most Bromeliad names are in Greek or Latin form, a scientific usage that is international. The Latin or Greek names usually refer to an outstanding characteristic of the plant, such as red leaves (*Guzmania sanguinea*), or to the person who discovered it (*Vriesea fosteriana*, after Mulford B. Foster), or to the geographical place where the plant was first found (*Aechmea orlandiana*). The genus names *Tillandsia* and *Billbergia* honor two Swedish botanists.

The Latinized names are always given in italic type, with the genus name capitalized and the species all-lowercase; after the first mention, the genus may be designated by its capital letter only; for example, *A. orlandiana*.

In addition to species, plants have been hybridized to produce outstanding flowers or unusual color. These hybrids are called *cultivars* and are denoted by single quotation marks, for example, *Aechmea* 'Meteor', which is known for its brilliant flower bracts. However, as work is still going on with the hybrids, the same cultivars may bear different names in different regions.

Bromeliads can also be known by common names. *Aechmea fasciata* is generally known as Grecian Urn worldwide, but its common name in some countries is Vase plant and in other regions, Silver Urn. Thus, common names are not always trustworthy; only the botanical name should be used when buying any plant.

In McIntosh's *The Book of the Garden*, published in 1838, one Bromeliad is mentioned—namely the pineapple.

Many species were also in cultivation in Berlin in 1857–60, but the Belgians were responsible for popularizing Bromeliads as houseplants. The first full treatise on Bromeliads as indoor plants appeared in *Die Familie der Bromeliaceen* by Joseph Georg Beer, published in 1857. From 1851 to 1885, Charles and Édouard Morren published the journal *La Belgique Horticole* containing colored plates of plants. Notable work was also done with Bromeliads in the latter part of the nineteenth century by C. Koch in Germany and E. von Regel in Russia.

In 1935 the plant family acquired increased recognition with the publication of *Bromeliaceae* by Carl Mez, a German botanist. At about the same time the vogue for Bromeliads crossed the Atlantic when Mulford Foster of Orlando, Florida, started growing the plants. His many works on the subject furthered Bromeliad growing in America, and many plants bear his name, including *Aechmea fosteriana* (already mentioned), *Dykia fosteriana*, and *Cryptanthus fosteriana*. This noted collector, grower, and writer founded the Bromeliad Society in 1950.

Natural Habitat

Although most Bromeliads grow from the state of Virginia southward to southern Argentina, the largest number of species are found in Mexico, Ecuador, Costa Rica, Brazil, Peru, and Colombia. In these geographical places Bromeliads literally adorn trees and branches with their jewel-like beauty, and also grow on rocks and some in the ground. Most Bromeliads are epiphytes (air plants) clinging to trees and rocks as hosts, but they are not parasites.

Bromeliads are generally tropical plants but altitudes in various countries determine just what species grows where. In the countries mentioned, plants may be in the ground at low altitudes and in the rain forests at high altitudes, and even in the mountains of Peru where temperatures many hit 40°F (4°C).

Mexico has many layers of ecological variation, and Bromeliads can be found in wooded regions, the evergreen tropical rain forests, the mountains, and finally in the cloud forests at about 6,000 feet. In each climate change—warm to moderately cool to cool—Bromeliads exist. For every 1,000 feet there is typically a climate variation of 3–7°F (1.7–3.9°C).

Bromeliads are herbaceous—or rarely arborescent and deciduous. In general they produce flowers only once. From then on, the plant enters into a process of decline that may last several years, until it dies. Its inflorescence is terminal—in rare cases also lateral—growing from the center of the plant on an extension of the extremely short stem. However, before the plant dies, it produces quantities of offshoots (called *keikis,* which is Hawaiian for "babies"). The offshoots develop from axillary buds. They may also develop some distance away from the parent plant at the end of underground rhizomes. The production of offshoots guarantees the perpetuation of the plants and contributes to the formation of the dense population clusters. Very special examples are some species of *Tillandsia, Orthophytum,* and *Ananas,* which produce offshoots at the base or apex of the inflorescence. In terrestrial Bromeliads, inflorescences are sometimes weighed down by the new plantlets as well as the actual fruits. Where they touch the soil new offsets may take root.

The intense coloration and variety of patterns on Bromeliad leaves endows them with great aesthetic value. Thanks to these color characteristics,

Bromeliads do not fade beside other typical flowering plant families. Inflorescences (bracts and flowers) take on a wide variety of shapes and sizes. They may be small and located deep within the rosette, below the level of the accumulated water, as happens in species of *Neoregelia*, or may appear like white bouquets, as in *Cryptanthus*. In other genera, they shoot vertically upward for several inches with large lateral branches. They may also be pendant, oblique, triangular, spherical, cylindrical, ellipsoid or corymbous; generously panicled or racemose, Bromeliads never cease to surprise. The colors range through the entire spectrum. Modified leaves take on the shape of bracts whose vivid hues decorate the scape (flowering stem), branches, and flowers. The colorization of the inflorescences may fade in a few days as in the case of certain *Billbergia*s, or last for months, as in some *Aechmea* and *Vriesea* species.

Anatomy of a Bromeliad

Most Bromeliads have a short central stem, and many Bromeliads have scales on their leaves; the scales act as a water-absorbing system. You can see the scales on Bromeliads like *Aechmea fasciata*—they form silver bands. Leaves differ—some are tough and firm (for example, *Aechmea*s and *Hechtia*s); several are edged with spines; such as some *Orthophytum*s or *Dykia*s. Others, such as the *Guzmania*s and *Vriesea*s, have leafy, spineless green foliage. Leaves may be plain and green or banded, mottled, streaked or spotted.

The inflorescences, which in most Bromeliads rise from the center of the plants, are tall and wiry; in certain plants, like *Neoregelia*s and *Nidularium*s, the flower is in rosette form and it is the color of the head (bract) that makes the plants so outstanding. The flowers are generally small— except for *Tillandsia cyanea*, which has 1-inch (2.54-cm) purple blooms. Many Bromeliads have brightly colored berries that remain on their stalks for months.

Bromeliads are generally tubular (*Billbergia*), vase (*Aechmea*), or rosette shaped (*Neoregelia*). The leaves form a cup or reservoir for water and food. In the Bromeliads' native habitat, small insects and organic matter drop into this reservoir, eventually dying there, furnishing nutrients for the plants. In cultivation, Bromeliads are largely free of such pests,

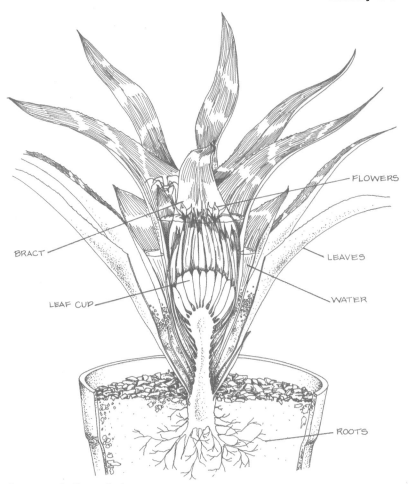

FLOWERS

BRACT

LEAVES

LEAF CUP

WATER

ROOTS

Anatomy of a Bromeliad

although a small frog has visited my *Aechmea fasciata* for more than a year, benefiting from a supply of fresh water and protection from predators. The cup or vase of Bromeliads should be filled with water and flushed with fresh water periodically—every week.

The leaves of many Bromeliads are in a spiral rosette pattern, packed close together, in a wide variety of colors, shapes, sizes and textures. A few species of *Pitcairnia* are deciduous, losing their leaves during the dry season. This group also includes *heterophyllous* species, in which the basal leaves of the plant are morphologically different from the upper leaves, such as *Pitcarnia heterophylla*.

Types of inflorescence

The leaf margins may have no thorns, or they may have small to very large thorns, in dense or sparse clumps. Generally, the base of the leaves (leaf sheath) is wider than the upper section, which is called the blade. The wider the sheath, the better the plant can hold rain water and organic litter. In these cases the leaf rosette is usually called a tank, vase, or cistern. However, not all Bromeliads have tank-type rosettes. When the sheaths are poorly developed, as is the case for the pineapple plant, no liquid accumulates. Some plants are *caulescent,* having an above-ground

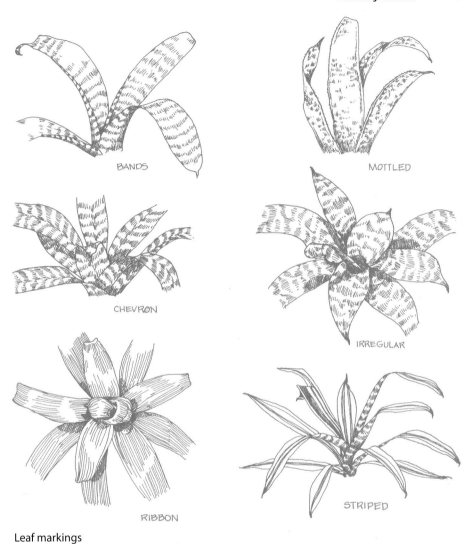

BANDS

MOTTLED

CHEVRON

IRREGULAR

RIBBON

STRIPED

Leaf markings

stem. Others feature leaves arranged distichously in a fan shape, as in some *Neoregelia*s.

Blooming Times

Most Bromeliads bear their colorful inflorescences in the spring and summer; it is then that my garden room is alive with vibrant colors. *Billbergia*s and *Aechmea*s predominate, and *Neoregelia*s and *Nidularium*s offer fiery crowns of color in the late summer.

In the winter several of my *Guzmania*s bloom, notably *G. monostachia* and *G. lingulata,* and some *Neoregelia*s are still colorful too. In fact there are few times during the year when there is not color in my garden room. Because flower crowns stay colorful so long (up to two months) for most Bromeliads it is difficult to specify blooming times as one might do for orchids. (Exceptions are *Tillandsia*s and *Billbergia*s whose flowers last only a few days.)

For Foliage and for Flowers

You can grow many species of Bromeliads strictly for their highly decorative foliage. For example, the beauty of *Vriesea hieroglyphica* or *Aechmea zebrina* cannot be surpassed. *Neoregelia*s are famous for their color at bloom time, and *Cryptanthus* species are highly prized for their leaf color. *Nidularium*s offer beautiful foliage because the symmetrical rosettes are so handsome. An outstanding plant, one I have treasured for years, is *Orthophytum navioides*, with dark green spiny leaves that are red at the base and shiny. *Vriesea fenestralis* and *V. hieroglyphica* have banded, colorful leaves, and not to be slighted are *Guzmania lindenii* and *G. zahnii.*

When we speak of "flowers" in Bromeliads, we mean the colorful bracts that retain their beauty for weeks. The actual flowers are small, protected by the bracts. In this group are the superlative *Aechmea fasciata* and *A. chantinii,* both examples of nature at its best. The former has tufted pink flower heads; the latter has brilliant red and yellow bracts. The *Billbergia*s, specifically *B. pyramidalis* and *B. zebrina,* have very large vividly colored bracts. The *Guzmania*s excel in flower crown color, including the red, white, and black *G. monostachia,* the long-lasting orange bracts of *G. lingulata,* and the charming yellow bracts of *G. zahnii. Streptocalyx* (once known as *Aechmea*) has bizarre but beautiful branched red inflorescence, and some of the *Quesnelia*s look like red-hot poker plants. *Tillandsia*s almost defy description. The brilliant red, candlelike inflorescence of *T. caulescens* and yellow waxy head of *T. fasciculata* especially deserve mention.

Now that your appetite for Bromeliads has been whetted by this brief overview, let's move on to the first course—getting started with cultivated Bromeliads.

Flower form

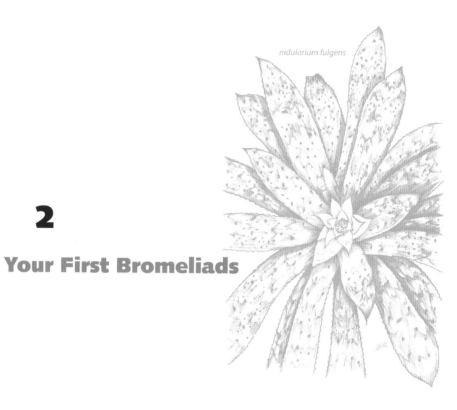
nidularium fulgens

2

Your First Bromeliads

Bromeliads make any nongardener look good because they retain their handsome appearance even if neglected. The water in the vases or cups provides a reservoir, plus a constant supply of additional nutrients from insect and leaf debris for the plants if you forget to water them. Like orchids, Bromeliads have an intense desire to survive, and most have tough, thick leaves that deter insects who choose leafier plants.

Of prime importance with having beautiful Bromeliads is how you treat the plants when you take them home from a supplier. In stores and nurseries, the plants may not always be maintained as they should be, so inspect your Bromeliads when you get them home to be sure they do not harbor insects such as mealybugs or aphids, for example. If you are concerned about insects in the pot, you can soak the container in a sink of water up to the rim. Any pesky critters generally come to the surface. Wash the above-ground portions of plants thoroughly and let them dry in a shady warm place. Inspect potting medium to be sure that it is fresh and not soggy or muddy. Also be sure your Bromeliads are in suitable pots with drainage holes in the bottom. Many genera, such as *Aechmea, Neoregelia,* and *Vriesea* for example, make fine pot plants. In containers, they need the usual water and fertilization you would give to most indoor plants; however, they do not do well if overfed (see chapter 3).

As with most plants, some species in a genus as well as some genera are favored over others—mainly because they are perhaps prettier or more leafy or have an attractive growth pattern. Among Bromeliads, *Aechmea*s, *Guzmania*s, and *Vriesea*s are the favored plants. The *Aechmea*s offer a wealth of beauty both in leaf and flower. *Guzmania*s, with their rosette or fan-shaped growth, are lush and always handsome, and *Vriesea*s have an infinite diversity in leaf pattern. Select from these genera and you are sure to be pleased with the family of Bromeliads.

When you get plants home do not put them in direct sun, which can shock the plant. Instead find a semi-sunny location and leave them there a few days before you move them into more light. Always put your plants in a area of good air circulation.

Most Bromeliads are nominally priced from about $10 and up, depending on the size and the rarity of the species. Newly introduced plants will cost more money, as will specimen (large) plants. If possible buy mature plants, although seedlings are available.

Do not try to bring in Bromeliads from foreign countries. Most countries do not allow any importation of Bromeliads unless you have suitable clearance permits from the authorities. Most Bromeliads and orchids are endangered species, so taking plants from their native habitats is not tolerated. Check local authorities in any country you are in if you decide to import Bromeliads.

Bromeliads are great for landscape use but also are equally desirable for home decoration. Because the Bromeliad family has so many different plants—small, medium, and large growers, ample variation is available to help bring color and nature to your interiors. They can be used as a special accent that affords the home owner use of color and more color.

Most *Aechmea*s, *Guzmania*s, *Neoregelia*s, *Nidularium*s and *Vriesea*s grow only to 40 inches (101 cm) tall, and many, such as *Cryptanthus* species and smaller *Aechmea*s can be used in confined areas because they only grow to about 6 to 8 inches (15 to 20 cm). With their varied shapes, the Bromeliads offer some sculptural accents in the home.

Basically, most Bromeliads like a bright place indoors but do not require full sun; rather, they grow best in a filtered light and also an area of good air circulation. So whether for a table or a desk there is a Bromeliad for every area. If you prefer a single accent to create drama, look to the *Hohenbergias* and *Streptocalyx* (*Aechmea*) species, which can grow to 40 inches (101 cm) or more and are best used as a single feature in a room.

Unlike orchids, which show only foliage when not in flower, the Bromeliad family is always attractive because of leaf color variation, such as *Vrieseas fennestralia* or *V. heierglyphica* with bars and zigzag designs on green leaves—a striking study in color.

While one may think of Bromeliads as living-room plants, they are also appropriate for table and desk areas, or especially in the bathroom where they lend a tropical aura to the room.

And best of all, Bromeliads need little care at home. Keep fresh water in the vase or cup of the plants, and most species grow on their own, always moderate to slow growth and never rampant growth, as in some other plants.

In Garden Room and Atrium

"Garden room" or "atrium" can be construed as screened porch, terrace, or patio. Bromeliads are perhaps the best suited plants for these partially open or completely open areas. In most cases plants can be grown in containers and moved at will if necessary. In the versatile family of Bromeliads, plants supply colorful flowers or beautiful foliage. Do not forget the sculptural effect of Bromeliads—some are statuesque, others fan shaped and some branching.

Just what you use depends greatly on the space. If it is a pool area, place plants at the edge of the space alongside one end of the pool, or spot them around the area in handsome containers. If the area is a patio, mass Bromeliads in containers in one place to create a dramatic effect. Fan-shaped or rosette-shaped plants, such as some *Neoregelias* or *Nidulariums*, can be used as colorful accents and are best placed at ground level for an effective scene. If a sculptural effect is wanted, select *Aechmea*s and *Vriesea*s, which offer a vase shape.

When you are using Bromeliads in a screened patio or garden room, add some leafy plants to compliment the picture. Large-leaved plants such as *Dracaena* blend well with Bromeliads.

Bromeliads make fine vertical accents against bare walls. *Billbergia*s, with their tall tubular growth, look handsome on low tables. *Aechmea*s and *Guzmania*s offer attractive whorls of color; *Vriesea*s and large *Tillandsia*s are graceful and palm-like. The flat fans of *Neoregelia*s and *Nidularium*s are fine for a dim corner or on a coffee table where color is appreciated. *Cryptanthus* species are attractive almost anywhere; a row of these small plants at a kitchen window is always pleasing. Contemporary rooms are enhanced by well-placed Bromeliads because both plant and flower form add dimension.

In lobbies and foyers of motels, hotels, and apartment buildings, indoor plantings are now almost essential. Artificial specimens leave much to be desired, and they gather dust. For these big areas, Bromeliads are the answer, even when they are grown at a northern exposure.

If you have an indoor garden, try some of these air plants among your green plants. A striped *Aechmea chantinii* or a plum-colored *Aechmea fulgens* var. *discolor* offers contrast and drama. If you have planter boxes, Bromeliads do better in them than almost any other plants.

Many Bromeliads—*Aechmea angustifolia*, for instance—have brilliant fruit that lasts two or three months. Then the color fades and the berries shrivel but remain. A unique effect can be achieved by spraying the berries with gold or silver paint. The metallic covering magnifies the depressions and markings of the dried berries, and they shine brilliantly as artificial light strikes the many facets. This look could be something unusual for a table in the entrance hall or in the dining room. The flower scape with berries and bracts can also be cut off and placed in an opaque vase (without water) to make an attractive study in line and texture.

How to Make a Bromeliad Tree Garden

Bromeliads adapt readily to tree branches or pieces of driftwood. To make a "tree," select a shallow round or rectangular container 1 to 4 inches (2 to

10 cm) deep so that it can support a branch about 24 inches (60 cm) tall. The idea is to make a miniature version of what might be seen in nature, where dead trees devoid of foliage look handsome with Bromeliads and orchids growing on them. Look for a branch that has interesting form. Anchor the wood in a nest of crumpled florist wire. Fill the dish with white pea gravel and top with sphagnum.

When I make a tree garden, I attach plants to the branch before setting it permanently in the container, but you can trim the tree after setting

Mounted Bromeliad

1. WET SPHAGNUM.

2. WIRE OR TIE WITH STRING TO TREE BRANCH.

3. PLACE PLANT AND WRAP IN SPHAGNUM. WIRE OR TIE IN PLACE WITH STRING.

4. WATER PLANT AND BARK.

the branch if you prefer. Either way, the Bromeliads must be firmly affixed. Cover roots with sphagnum moss and wire the clumps around the branch. Use plastic-covered wire from a floral supply house. *Do not use galvanized wire;* this causes leaf burn if it touches the foliage.

Miniature Rock Gardens and Terrariums

Bromeliads make attractive indoor rock gardens. They are one plant group that will survive in this situation where other plants usually succumb after a short time. A 2- or 3-inch (5- or 7-cm) deep aluminum baking dish or a metal box filled with gravel makes a good container. On the surface, put interesting small rocks or stones. Then arrange pots of Bromeliads on the gravel with the rocks as background. Small plants of *Cryptanthus* species and *Vriesea*s are handsome in these compositions.

A similar idea can be used to make a terrarium. Spread 2 inches (5 cm) of gravel-and-sand mixture over the bottom of a 10-gallon aquarium. Set in small stones and interesting pieces of dry branches and set potted Bromeliads (or plants without pots, with roots wrapped in sphagnum) to create a pleasing design. I arrange the greenery from the back corners of the tank to front center, larger species in the rear and smaller subjects forward.

I select *Cryptanthus* plants because I know they will thrive there. The highly colored foliage with wavy bands suggests starfish, and in the changing light and under the distortion of glass, the terrarium looks like an undersea garden. Tiny ferns or other dwarf plants can be added. Set the terrarium away from the sun, preferably at a north exposure close to the window, and cover with a pane of glass cut 2 inches (5 cm) shorter in the width than the top of the terrarium. This allows air to circulate inside, and the plants to remain healthy.

I do not really water plants in my terrarium. Instead, about twice a week I mist them with tepid water from a spray bottle (see chapter 6).

On Patios and Courts

Patio and outdoor living areas are now a way of life even where summers are short. For these areas, Bromeliads are ideal. In climates where it is safe to have plants outside from June to September, leave the Bromeliads in pots and, when weather permits, arrange them outdoors in decorative compositions. A rock or large tree makes an effective background.

Just as Bromeliads enhance the in-home and patio environments, they are also a natural choice for landscaping, as you'll see next.

3

Bromeliads in the Landscape

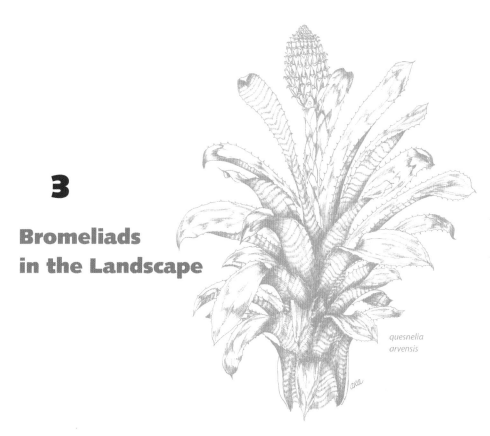

quesnelia
arvensis

Bromeliads in a temperate climate have a myriad of uses. They are desirable in the landscape as barriers or as accent-islands, or simply placed in an ornamental container. And, because so many species and varieties are available, the choice is vast.

Roberto Burle Marx of Brazil was perhaps the first to realize the beauty and usefulness of Bromeliads, placing plants in his native country as a fence or barrier. *Aechmeas*, *Bromelias*, and other Bromeliads were put to use. He also used Bromeliads as garden accents selecting the colorful *Neoregelias* and *Nidulariums* as color backgrounds and in groups as structural parts of the garden design.

In subtropical climates such as Florida, Bromeliads have many uses. Indeed, once installed they form a unique garden departure from other plants such as hibiscus and orchids. While some Bromeliads are terrestrial, many are air plants (epiphytic), such as *Tillandsia*s, that can be grown attached to tree trunks and branches to create color in your yard.

Especially around pools, Bromeliads in containers add that touch of color to make the area attractive. The variation and color of leaf patterns also make them desirable and pleasing throughout an atrium or screened area, as mentioned in chapter 2.

In some warm climates, Bromeliads can *be* the garden, because there are many to choose from. The rosette- or fan-shaped plants, such as in the genera *Neoregelia* and *Nidularium,* blaze with color and can set the stage for the garden. Vertical or tubular Bromeliads, such as *Billbergia*s and *Vriesea*s, add a vertical accent to the yard, and the variation in leaf patterns in many Bromeliads makes them ideal for gardens.

Year-Round Care

When caring for Bromeliads, the amount of water and light are the primary variants that need to be considered in the course of the year. Bromeliads do best in climates where winter temperature remain mild, rarely dropping below 40°F (4°C).

Spring

Keep plants somewhat dry at the roots but keep the cups or vases filled with fresh water. Generally bright light suits the plants at this time of year. Do routine grooming now, such as clipping dead leaves and being sure plants are propped and upright. Some Bromeliads are heavy and have a tendency to tip, so be aware. This is especially true of plants in plastic pots.

Be sure plants are in an area of good air circulation. Stagnant air locations should be avoided.

Most Bromeliads react poorly to excessive feeding so be cautious and feed only a few times in spring with a mild 10-10-5 fertilizer solution. (Follow directions on the bottle.)

Summer

Keep the potting media evenly moist, and place the plants in a somewhat sunny location so leaf color can develop. Feed every week now with a mild solution as mentioned above. If a plant is not doing well in one area, move it to another place. Sometimes moving only a few centimeters will suit a plant better than its previous location.

Fall

It is time to lessen watering now; apply moisture about once a week to the medium, and keep the cups or vases filled with fresh water.

Winter

Gradually reduce watering and generally groom and trim the plants. Also do any repotting from now until spring.

Choosing Containers

Today at most garden shops you can find an array of containers in every shape, size, material, and color. I have in the past always opted for terra-cotta pots, but lately manufacturers have responded with so many handsome containers that there is always at least one type that suits our home décor.

Bromeliads do fine in containers or pots, but be wary of the plastic containers. *Aechmea*s, *Billbergia*s, and *Vriesea*s, with their vase shaped growth, can be somewhat top heavy and in plastic pots are liable to tip over. Be sure containers have drainage holes at the bottom so excess water can escape. Glazed pots also should be avoided because the glaze prevents moisture from evaporating through the walls, which can create a soggy medium detrimental to plants.

If you decide to use decorative jardinières, which are very attractive, put the plant and its container in the jardinière, using its receptacles more as coveralls, rather than planting directly in the containers (also see chapter 6).

In the following chapter, you'll learn more of the details of Bromeliad care.

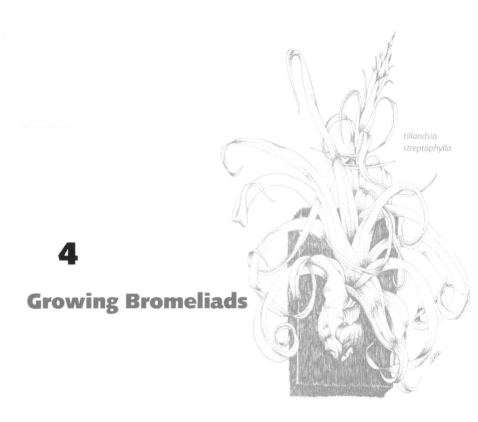

*tillandsia
streptophylla*

4

Growing Bromeliads

Bromeliads are amenable plants that can, if necessary, take abuse. Because most have their own watering "cups" or "vases" that act as reservoirs, they can grow almost on their own. As mentioned, if you flush the cups with fresh water every week they will remain free of dirt or dead insects. And if rains are frequent you need not worry about watering; nature does the job for you.

Warm days and sunshine are what most Bromeliads thrive in. At night if temperatures drop somewhat, say to 50°F (10°C), no harm will be done to Bromeliads—and they can even take temperatures down to 40°F (4°C). Frequent summer showers suit these plants, and although sun enhances color, it is best to place them where they receive dappled sunlight—perhaps under a tree or in a similar area.

Occasionally a plant looks dry. The appearance of the foliage and the whole attitude of the plant looks wan. If plants are in the landscape remember that leaves and debris decompose in the vases of the plants, providing nutrients to the plant. If plants are under a roof, provide mild fertilization using a weak 10-10-5 fertilizer a few times in the summer and once or twice in winter. Too much fertilizer can do more harm than good with Bromeliads.

Bromeliads, like orchids, require good air circulation to grow well. Nature generally does the job for you. On stagnant days with high humidity, both people and plants learn to tolerate the temperature changes of nature. Books on Bromeliads recommend that readers put some Bromeliads in sun, but I have found that dappled light works best, and while some of the Bromeliads do need full sun, most do not (see the Quick Reference Guide at the end of the book).

Planting and Potting

Terrestrial Bromeliads can be grown in various media. Some plants, such as *Aechmea*s, can be grown in fir bark, and some gardeners use a combination of both soil and bark. *Tillandsia*s, which are mostly air plants, are planted in fir bark, the accepted choice, or better, mounted on tree branches.

The potting procedure with most Bromeliads is the same as potting any plant. Basically I put drainage shards in the bottom of a clean container, preferably terra-cotta. As noted in earlier chapters, plastic is sometimes too light to provide a firm base. Fill the pot half way with bark and soil. Center the plant in the medium and tamp down with your thumbs or a potting stick. Most Bromeliads thrive in tight potting. Remember to stake tall plants with sticks if necessary and label with identification tags. Water lightly and place in a bright light.

Keep the plant in bright light but not direct sun, scantily watered until the plant takes hold, and then place it in dappled sun and water routinely, as described elsewhere.

Removing a Bromeliad from a pot is simple. Rap the bottom of the pot on the side of an old table or floor to loosen the root ball, then grasp the Bromeliad at the base with your hand and gently tug. Wiggle it until it comes out of the pot, being careful not to injure the roots. Don't pull it

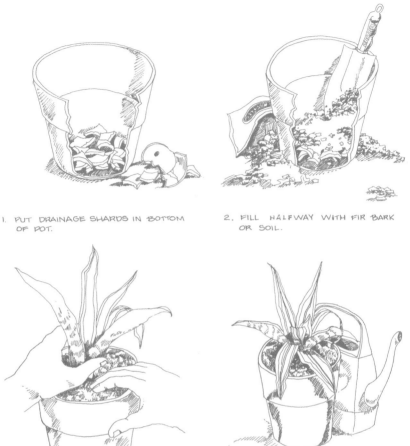

1. PUT DRAINAGE SHARDS IN BOTTOM OF POT.

2. FILL HALFWAY WITH FIR BARK OR SOIL.

3. CENTER THE PLANT. FILL IN WITH SOIL, BARK, OR COMBINATION OF BOTH. TAMP DOWN.

4. WATER LIGHTLY AND PLACE IN BRIGHT LIGHT.

Potting with bark or soil

hard or force it loose. When it is out of the old container, shake off the old medium or crumble it off with your fingers.

After potting or repotting always place the plant in a dim light and out of direct sun. Wait a week or so and water scantily, then, once the plant shows signs of perking up, move it into dappled light or sun and water routinely.

2. TURN PLANT ON SIDE AND CUT IN HALF.

1. REMOVE PLANT FROM POT. TAP EDGES AGAINST TABLE.

3. SEPARATE INTO TWO OR THREE SMALL CLUMPS.

4. PLACE EACH NEW PLANT IN SEPARATE POT.

Division

Climate and Bromeliads

Many variables apply to climate changes and Bromeliads. If temperatures do fall below 40°F (4°C) for more than a few nights, protect your plants. Simply throw a sheet or newspapers over the plants. If you use a plastic shield be sure there are air holes in the sheet. It is also important that you protect your plants from a hard freeze and have their cups or vases filled with water to insulate the plant parts. To sum up temperature and Bromeliads, if you are comfortable, your plants will be also.

Cold Hardiness in Bromeliads

Growing Bromeliads in various climatic zones is a much-discussed topic of conversation with hobbyists. But basically there are no hard and fast rules about temperatures and Bromeliads. The climatic charts made by scientific data cannot hold true for all Bromeliads because of the wide variance of temperatures and microclimates. But generally, I can make the following suggestions for some plants: *Aechmea recurvata, A. ornatea,* and *A. cylindrata* can tolerate temperatures to 50°F (10°C), as can *Billergia nutans* and hybrids, *Dykia* species and *Neoregelia*, some *Puya* species, *Quesnalia,* and *Vriesea mariae.*

Some of the semi-hardy Bromeliads that can tolerate temperatures to 35°F (1°C) are *Aechmea* 'Burgundy,' *A. gamosepala, A. calyculata, Billbergia pyramidalis,* several *Neoregelia* hybrids, *Guzmania wittmackii,* and many of the hybrid *Vrieseas.*

I found that most of the Bromeliads I cultivated did well in average temperatures of 60–75°F (15–23°C) or a few degrees one way or the other say to 78°F (25°C). In general I can say that most Bromeliads can tolerate a varying temperature difference of say 5–10°F (2–5°C).

Further, although scientific charts and data are available, as mentioned, on cold hardiness in Bromeliads, one cannot say that these temperatures are accurate because controlled experiments, to my knowledge, have not been done. Thus, feel free to grow Bromeliads and enjoy them.

Propagation

Bromeliads are very free with their offshoots, and creating new plants from old is relatively easy. Most Bromeliads produce more offshoots and suckers than other plants. This botanical feature makes up for the fact that Bromeliads only bloom once.

The growth habit of *Aechmeas* differs from other genera, and "pups" or "keikis" from the mother plant live on for some months. *Neoregelias* develop offshoots close to the plant, and the offshoots of some *Vrieseas* start in the cup close to the flower stem. When offshoots are about 3 to 4 inches (7 to 10 cm) tall they can be clipped from the plant or twisted off and started in fresh pots of medium to bear another plant. Most Bromeliads

1. WAIT UNTIL SHOOT IS 3" TALL.

2. SEVER WITH SHARP KNIFE.

3. POT IN BARK OR SOIL.

4. PUT IN PLASTIC BAG FOR A FEW DAYS.

Offshoot propagation

produce offshoots after the plant blooms; others start their offshoots before the mother plant comes into flower.

It is easy to see the new growth when it starts, and whether to leave them on the plant for a month or so to grow strong or take them off when they first start growth is up to you. The longer you leave the offshoots on the plant, the more offshoots you may get. I wait to remove them until I see a few leaves.

In summary, getting new plants from old Bromeliads is a simple matter, and almost anyone can perform this procedure.

Forcing Flowers

Some Bromeliads may be reluctant to bloom even if the culture conditions are good. If a mature plant has not bloomed, you can stimulate flowering by giving the plant ethylene gas. Use ripe apples, which emit this gas. Put the entire plant, pot and all, in a clear plastic sack with a ripe apple. Be sure to empty the cup of the plant because excess moisture while the plant is in a closed bag can create rot problems. Tie the bag and place it in bright light (but no direct sun) for about seven days. Then remove the bag and dispose of the apple. The plant should bloom in about a month, depending on the species.

Grooming and Trimming

All plants need some grooming to keep them attractive, and Bromeliads are no exception. Leaves at the base of plants naturally die off and should be removed before they become diseased and rot. Leaves that have turned brown at the tips have usually been bruised—perhaps the plant was pushed against a window. Trim the leaves with small, sterile scissors. Keep Bromeliad leaves clean and shiny—dust and soot settle on leaves and clog pores, so occasionally wipe foliage with a damp cloth.

Most Bromeliads should grow erect. If a plant begins to topple, firm the potting medium around the base of the plant. You should stake tall tubular species to eliminate the possibility of plants toppling over. To stake a plant, insert a small stick at one side and plunge it into the soil, then tie the plant at its center to the stick. Use string for tying—do not use wire ties or hard plastic or you may cut the plant.

Remove faded flowers and bracts after plants bloom by cutting them off or by pinching them off with your fingers.

An important part of Bromeliad maintenance can be the control of pests; see the next chapter.

5

Pest Problems

vriesea
schwackeana

Bromeliads are not an insect's favorite dinner. Most Bromeliad leaves are leathery, too tough for "critters" to penetrate. Tightly packed plants create a breeding ground for insects moving from one plant to another, so give Bromeliads space and provide good air circulation. While typical insect pests such as aphids and mealybugs are not particularly prevalent among Bromeliads, scale insects, both the black and white varieties, cause havoc, leaving a waxy white coating that retards growth.

Common Problems

If you encounter scale on your Bromeliads, use an insecticidal soap or a liquid detergent on a cloth, and afterward sponge leaves with clear water. I also have thwarted scale by applying rubbing alcohol on cotton swabs and then washing leaves with tepid water. Be sure to wash away the culprits to also get rid of the eggs.

A recent problem with Bromeliads, especially *Tillandsia*s, is a beast called *Metamasius callizona,* a weevil-like culprit, whose larvae generally attack the base of the plant. Research is ongoing to develop a biological control on this weevil.

Thrips are needlelike insects that cause grayish foliage, and whiteflies are tiny pests that cause speckled foliage.

Look for insects in the axils of the leaves and the undersides of leaves, where they are most prone to attack, and take immediate action to control them before they have a foothold.

And, as mentioned in earlier chapters, by flushing the tanks or vases of the plant with water about once a week, you'll help prevent the reservoir from becoming habitat for mosquito larvae and generally keep it clean of debris.

Preventives

If you observe your plants every few days and catch trouble before it starts, no high-powered insecticides are needed, nor should they be used. Old-fashioned remedies such as laundry soap and water eliminate many culprits, and alcohol, as mentioned, also deters most pests. Tobacco steeped in water a few days and applied directly to insects will get rid of them quickly, and a simple heavy spray of water over the entire plant always helps to keep plants healthy.

If Bromeliads are well cared for, they rarely develop fungus or bacterial diseases. Once again, good ventilation will deter many diseases. Disease shows up in spots, mildew, rusts, and rot. Fungi, like mildew and rot, enter the plant through cuts or wounds in the plants. Their spores are carried by wind, water, insects, and contaminated equipment. Check with local nurseries for safe remedies.

6

Gallery of Bromeliads

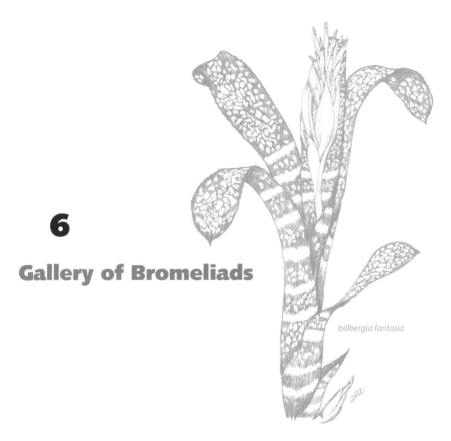

billbergia fantasia

In this chapter you'll find an appealing menu of many popular varieties of Bromeliads from 20 different genera. These summary descriptions and images should help you choose plants you may wish to grow. The genera are in alphabetical order.

Aechmea

The genus *Aechmea*, with more than 200 species, is the best known of the Bromeliads. They are readily available and have beautifully colored foliage. Also, these plants produce flower crowns that last for months; recent hybrids have put these Bromeliads at the front of the pack, with each new plant seeming more handsome than its predecessors.

Plants range in size from 24 inches (60 cm) to about 4 feet (122 cm) in diameter, and their shape is usually a tubular rosette—almost a funnel type of growth. The leaves, edged with tiny spines, may be plain green, but often are green banded with handsome silver, maroon, rose, or purplish brown. Some species have leaves blotched or striped with color rather than banded, and a few have leaves with a combination of all three markings. The inflorescence may be branched, cylindrical, or pendant (as in *Aechmea racinae*). The flowers themselves are usually small; the bracts are responsible for the vivid colors and last for weeks. Most *Aechmea*s bear berrylike, brilliantly colored fruit after they flower. As a rule, these plants are very free with offshoots, frequently throwing three or four keikis from a mature shoot.

The plants are native to central Mexico and south to Argentina, and most grow on trees or rocks while some species inhabit the forest floor. It is not uncommon to find them at 6,000 ft. in bright sunlight or at sea level in dappled shade. Generally, the tree-dwellers outnumber the terrestrials.

*Aechmea*s under cultivation grow best in bright light. I pot my plants in equal parts of fir bark and soil; this medium offers excellent drainage. I keep the growing medium evenly moist all year, and I do not worry about watering the plants regularly. They can survive a week without water. However, water should be present in the vase of the plant. If the vase is left dry for too long, plants suffer.

A sunny location is best, but some of *Aechmea*s do well in indirect light too, although their foliage will not be as brilliantly colored. Most bloom without pampering if light and moisture are good. The smooth-leaved *Aechmea*s can be wiped with a damp cloth occasionally, but those with scaly leaves (and a handsome powdery coat) should not be handled because they get marred when touched. Most plants can be misted with tepid water to keep humidity high. It is essential that *Aechmea*s be in a location with excellent air circulation or they will not do well.

During my time of growing *Aechmea*s I have rarely seen any insects on the plants; most are generally free of pests and disease as well, making them stellar houseplants. One disadvantage is that tall species, such as *A. luddemanniana* and *A. chantinii,* become top heavy when filled with water and can fall over, even when planted in clay pots. Also, the base of the plant may become weak and bend; I tie tall *Aechmea*s to small sticks with string.

Recently plants of the genus *Streptocalyx* have been included by some as belonging to *Aechmea;* see the entries for that genus, later on.

1. *A. blanchetiana* (*fraudulosa*)

Generally tree-growing, from Brazil. An open rosette 3 or 4 feet across; orange-golden leaves. Grows well in partial sun. Bright red influorescence. Best for landscape use.

2. *A. retusa*

A thin, tubular plant that grows to 14 inches (35 cm) with grayish green leaves spotted maroon; spines on edges; lovely orange flower bracts, yellow flowers. Urn-shaped, from Brazil and Colombia.

3. *A. caudata* var. *variegata*

Growing to 36 inches (91 cm) tall, this plant would be a good investment for its handsome foliage even if it never flowered. The leaves are yellow-white and green striped. The flower head, yellow-orange, erect, and striking, usually appears in winter. This is one of the few *Aechmea*s difficult to cultivate, but a healthy specimen is worth the effort.

4. *A. chantinii*

Grows to 36 inches (91 cm), this queen of the *Aechmea*s is indeed handsome. Leaves are olive green with silver bands. The large flower head is upright, a vivid red and yellow. Of medium size, this regal plant gives weeks of pleasure when it blooms in spring.

5. *A. chantinii* 'Silver Ghost'

An upright, 40-inch (101-cm) rosette of handsome, pale silvery green leaves banded with darker green. Spike erect and branched, pale rose bracts and greenish yellow flowers.

6. *A. fasciata*

Sometime called the urn plant, grows to 24 inches (61 cm) tall and is a window gardener's delight. It produces a large pink inflorescence with dozens of tiny blue flowers in summer. Fine for the beginner, sure to bloom.

7. *A.* 'Foster's Favorite'

Plant grows to 24 inches (61 cm) with wine-red foliage and dark-blue winter flowers followed by red berries. A small, inexpensive air plant readily accommodated at a window.

8. *A. fulgens*

Plant is only 20 inches (51 cm) tall with maroon-shaded green leaves. In spring, it produces dark-purple flowers followed by rose-colored berries that last for months. It is well suited to a limited space and will survive considerable neglect.

9. *A. gamosepala*

Introduced in 1890 from Brazil. Medium sized with glossy green leaves, urn shaped. Dense blue flowers.

10. *A.* 'Moorei'

A very showy Bromeliad, to 28 inches (71 cm); has an inflorescence similar to *A. chantinii* but with lower bracts carmine rose and upper flattened bracts lime green tipped with flame orange. Foliage is green.

11. *A. nidularioides*

Generally epiphytic and grows best in heat and humidity. Reddish green leaves. Cone shaped inflorescence rising to about 10 inches (25 cm) from urn-like growth.

12. *A. nudicaulis* 'Variegata'

A very handsome plant, makes a 30-inch (76-cm) tubular vase of bright green leaves lightly lined with yellow, leaf edges spiny. Floral bracts bright rose, yellow flowers.

13. *A. orlandiana* 'Burning Ember'

A showy 18-inch (46-cm) rosette of bright yellow-green leaves with brown cross-banding and heavy black spines. The arching spike has salmon-colored bracts and ivory flowers. Needs more heat than most in the genus. Bright light makes leaves colorful.

14. *A. racinae*

Grows to 14 inches (35 cm) and has light-green leaves and yellow-and-black flowers on a pendant spike. The orange-red berries appear appropriately at Christmas time and last until April. Try this one in a hanging basket and grow it in light shade.

15. *A. recurvata*

A tall 20-inch (51-cm) plant with thorny cactus-like foliage and a low pink flower head. Grow it on the dry side and in sun.

16. *A. woronowii*

A 20- to 30-inch (66- to 76-cm) plant with upright vase shape. Foliage bronzy green. In-florescence orange or pink.

Ananas

Ananas is one of the oldest known group of Bromeliads. It includes the pineapple, which has been widely grown for several centuries. In the fifteenth century, the pineapple was a known crop in the New World. As houseplants, several species have their merits, but they are not really exceptional because they simply are not as decorative as other Bromeliads. The miniature variety seems to have found a place in American homes more as an oddity than as a good indoor subject.

About ten species of *Ananas* are known, all terrestrial, and native to Brazil, Paraguay, and Venezuela. They have been successfully cultivated in Hawaii for many years. Most species have large rosettes—to about 48 inches (122 cm) across—and thus demand space. Leaves are either plain green or, in the decorative species, striped yellow and green, and are armed with spines.

The fruiting head develops on top of a stout stem that comes from the center of the plant; the head bears a small rosette of leaves that is a miniature version of the parent plant. This head can be cut and started as a new plant. The flowers have purple-blue petals. *A. nanus* is a fine houseplant because it is small. Plants require good sunlight to grow well but can tolerate warmth or coolness (55°F, or 13°C).

17. *A. comosus* 'Variegatus'

The commercial pineapple plant, growing to 30 inches (76 cm) tall. The stiff, spiny-edged leaves are gray-green and the purple flowers are produced at the top of an erect stalk.

18. *A. nanus*

A 10-inch (25-cm) dwarf species with satiny green leaves 10- to 14-inches (25- to 35-cm) long. It is available at florist shops and somewhat resembles *A. comosus*. Bears a small pineapple and is decorative on its stiff stalk. A good table plant.

19. *A. bracteatus* var. *tricolor*

Handsome plant with variegated leaves. Open rosette. Excellent for landscape. Colorful bracts, small flowers.

Billbergia

Some good plants can be found in this group as well as some that are not so attractive. *Billbergia nutans* varieties take up space, and the plain green foliage is not very striking. However, these plants do bloom profusely— sometimes three times a year. Most *Billbergia*s are native to eastern Brazil; however, a few come from Mexico and Central America. In nature the plants grow best in an arboreal existence. The plants also grow at ground level.

*Billbergia*s have fewer leaves than most Bromeliads—five to eight. They are easily distinguished from other plants because they are definitely tall and tubular; they do not have the rosette shape of the *Aechmea*s or *Guzmania*s. Leaves can be plain green, but more often they are mottled, banded, or speckled with contrasting color. Most *Billbergia*s have brilliantly colorful bracts (pink or rose), and as a group they are more floriferous and easier to bring into bloom than most Bromeliads. The flowers are generally small, of vivid purple, blue, and green. The inflorescence is pendant and short-lived, perhaps three to five days.

*Billbergia*s grow almost by themselves and need very little attention. If situated in a bright place and given even minimal watering, they grow— and fast. In fact, they grow so quickly that they need frequent repotting, which can become bothersome. Grow in equal parts sand, fir bark, and soil. Drainage must be perfect, and the growing medium should be kept just moist, with the vase filled often with water. Mist plants with tepid water frequently, especially during hot weather.

Bright locations produce very colorful foliage; in shade, foliage is somewhat dull. Even in low light the plants survive for a long time.

These plants are favorites in the garden in all subtropical and tropical regions, where they will naturalize on trees and create handsome scenes. *Billbergia*s produce offsets easily, or you can divide plants by slicing a crown with a sterile knife. *Billbergia*s are not as demanding as most Bromeliads in terms of air circulation.

20. *B. amoena*

Plant grows to 28 inches (71 cm), tubular with green leaves and rose-colored bracts. Flowers are green with blue edges.

21. *B. elegans*

Has a vase-like rosette with broad, light-green leaves margined with brown spines. It grows to 16 inches (40 cm) with pendulous inflorescence with rose bracts and yellow-green flowers. Takes coolness to 50°F (10°C).

22. B. eloiseae

Handsome species; silver-green leaves. Likes warmth. Tubular. Good outdoors.

23. B. 'Fantasia'

A 30-inch (76-cm) tall hybrid that is a perfect size for the windowsill, with green leaves spotted ivory, green, or rose. The pendant inflorescence of scarlet bracts and blue flowers makes it most appealing.

24. B. leptopoda

Has broad green leaves, spotted cream color. It rarely grows above 12 inches (30 cm). Leaf tips curl and the inflorescence has pink bracts and green-and-blue petals. A good pot plant.

25. B. lietzei

Only 12 inches (30 cm) tall, it is a charming small plant that lights your window with bright cerise flowers around the holiday season.

26. *B. pallidiflora*

Found in Mexico at about 4,000 feet. Tubular shape. Reddish gray leaves with silver bands. Arching inflorescence with pink bracts and green petals.

27. *B. pyramidalis*

A favorite with collectors. It is a broad-leaved 24-inch (61-cm) bottle-shaped plant, golden green, the compact flower head densely set with pink to red flowers on a short scape. A very beautiful *Billbergia* that blooms in winter. *B. pyramidalis* var. *striata* combines beautiful flowers with pale green-and-white foliage.

28. *B. pyramidalis* 'Kyoto'

Open tubular growth. Handsome foliage; green with white-banded leaves.

29. *B.* 'Strawberry Marble'

Extremely colorful foliage with red-and-ivory leaves. Great outdoor accent.

Bromelia

Members of this group are rarely seen in plant collections, and yet there are two species that are extremely handsome: *Bromelia balansae* and the dwarf *B. humilis*. They are perfect houseplants, and they make a dramatic statement in bloom because the entire center of the plant turns fiery red, like a volcano.

There are actually about fifty species of *Bromelia*s. The plants are native mainly to Brazil, Argentina, and Paraguay but some are found in Mexico, Guatemala, and the Caribbean. In nature *Bromelia*s grow in thickets on the ground and are often used in place of hedges to define property and keep out intruders.

Most are very large plants (with the exception of *Bromelia humilis*), up to 4 feet (122 cm) or more across, growing in a rosette shape, with plain green leaves edged with sharp spines—difficult to handle as a houseplant, but no more so than cacti. A few *Bromelia*s have variegated foliage. When potting or repotting *Bromelia*s, handle them with newspapers or burlap to avoid being scratched. If you are scratched by the spines, do treat the wounds or they may become infected. The plants may do better in ample space than in containers.

Indoors, *Bromelia*s have little trouble adjusting to varying conditions and so can be grown in warmth or coolness (to 50°F, or 10°C) in bright light or in shade. They can be grown in large-grade fir bark with no soil. Keep drainage perfect, and water plants frequently and heavily. I try to give my plants a sunny place because then the foliage becomes highly colored. Good direct sun is necessary if you want plants to produce their fiery display in midsummer. The plants produce many offsets. *Bromelia* foliage is so tough that insects rarely come near them, and I have never had any disease strike my plants. As decorative plants *Bromelia*s are especially good.

30. *B. plumieri* (Karatas)

A relatively recent intro-
duction (1968) from the
West Indies. Rosette shape.
Large and with spine-
edged leaves.

31. *B. serra* 'Variegata'

From Argentina. Smaller version of *B. balanse*. Leaves edged with white or pink.
Handsome.

Canistrum (Wittrockia)

I had been growing Bromeliads for several years when I visited the Fantastic Gardens in Miami, Florida, in 1969 and saw my first *Canistrum*s. To me the plants resembled *Neoregelia*s and *Nidularium*s but were somewhat more upright in growth and a little smaller. They make excellent houseplants.

Canistrum is a small genus of only twelve species native to Brazil. They have rosette-type growth and pale green leaves usually mottled with darker green. The white-and-green flowers are small and the flower head compact, hidden in the center of the plant surrounded by colorful bracts. In nature *Canistrum*s grow best in arboreal locations, usually growing on a tree limb or on rocks. Plants vary from 20 to 30 inches (51 to 76 cm) across.

*Canistrum*s grow well in a bright place. Sun is not absolutely necessary; they seem to do better in dappled light. They like moisture and their leaves clean—wipe with a damp cloth; provide the humidity at about 30 percent. My plants have not been as free with offsets as most other Bromeliads and generally have been more difficult to bring to bloom.

32. *C.* (*Wittrockia*) *gigantea*

Open rosette with broad shiny green leaves spotted purple. Handsome decoration.

33. *C.* (*Edmundoa*) *lindenii* var. *marginata*

A 30-inch (76-cm) rosette with exquisite yellow-and-green banded leaves. Rose-shaped inflorescence has whitish green bracts.

Cryptanthus

Cryptanthus species are foliage plants—the inflorescence is insignificant, but the leaves are exquisite and as handsome as the popular Rex begonias and much easier to grow. A window can accommodate as many as twelve to fifteen plants because most *Cryptanthus* species are small, growing to only 18 inches (45 cm) across. As houseplants they are amenable and adjust to conditions, but it takes some time before they grow readily.

About fifty species make up the group, along with dozens of hybrids. These terrestrial plants come from Brazil, and in nature they grow under varied conditions, most in shade; but in cultivation they prefer a warm and bright location. Most species are low-spreading rosettes, with about ten to twelve 2- to 12-inch (5- to 30-cm) leaves per plant. The leaves are somewhat crinkled and generally mottled and striped with unusual colors: brown, rose, silver, chartreuse, copper, pink, and so on. All species have tiny white flowers in the center of the plant. Offshoots are produced from axillary buds.

As a group the members of *Cryptanthus* have puzzled me. I have tried various ways of growing them but have not found any one way better than another. At one time I grew them in a terrarium; another time I wired them to a large tree branch and put them in direct sun at a window. Still another time I used them as decorative plants in small matching containers on my desk under artificial light. Either my test periods were not long enough or I did not have a proper control group, but I did not find that plants responded better one way or another—that is, with more colorful leaves or more flowers. I decided to grow *Cryptanthus* plants on a tree branch, and that is where they have been for more than ten years. The roots are embedded in years-old fir bark, and the plants look attractive. The branch is suspended from a wire in the ceiling so plants are at eye level.

You can also grow the plants in standard houseplant soil in 2- to 4-inch (5- to 10-cm) deep pots. Growth is slow. Insects rarely bother the plants. I recommend these Bromeliads if you have limited space and want something other than philodendrons or grape ivy at home. Keep humidity at about 30 percent, and mist foliage occasionally.

34. *C. acaulis* hybrids

Grows to about 5 inches (12 cm) across. The pointed leaves are apple green with a slight gray overcast. Tiny white flowers hide deep in the center of the plant.

35. *C.* 'Cascade'

Crinkled curled leaves. Small rosette.

36. *C. fosterianus* 'Elaine'

A hybrid with bronzy brown leaves. Most popular.

37. *Cryptanthus* in terrarium

Various species and hybrids—all with colorful leaves—make a handsome natural scene.

Dykia

*Dykia*s, which resemble cactus, are ideal houseplants for the beginning gardener because they are very adaptable to most growing conditions and can, if necessary, withstand drought. The plants range in size from 6 inches (15 cm) to about 24 inches (61 cm) in diameter and have stiff, spiny-edged leaves. The smaller species make handsome desk and table accents in decorative pots.

Dykias grow in fir bark and in soil. They seem to do best in bark. From central Brazil, these rock-growing xerophytes grow in full sun in rocky crevices. The plants need excellent drainage to thrive—any waterlogged medium harms plants. Grow them somewhat dry at all times. They have excellent leaf coloring in sun, but in shade foliage color is less pronounced.

Because of the spines, plants are difficult to handle; wear gloves. Once potted, *Dykias* do not need repotting for several years.

38. *D. fosteriana*

The most popular species and rightly so. It makes a handsome 12-inch (30-cm) plant with silvery purple rosettes that cascade over the pot; recurved spines. Large orange flowers in spring or summer.

39. *D.* hybrids

Thorny gray leaved; small rosette; dwarf. Good in planting beds.

40. *D. microcalyx*

Thin leaves; gray; open rosette.

Guzmania

These small- to large-sized Bromeliads almost outrank *Aechmea*s as the best Bromeliads, and rightly so because an incredible variation of leaf color and pattern is available. The decorative-leaved beauties, such as *G. lindenii,* provide startling accent, and even seedling plants are colorful.

The genus *Guzmania,* with about 200 species, grows in the forests of Ecuador and Colombia; other species come from Central America, Brazil, Costa Rica, Florida, and Panama. The plants grow in varying elevations, and it is not unusual to find *Guzmania*s at over 6,000 feet where nights get very cold. Generally, *Guzmania*s are in shady situations on tree limbs but protected from direct sun. They grow on lower limbs rather than treetops. They are epiphytic, but a few also do grow as terrestrials in nature. Plants have smooth-edged, glossy green leaves or leaves with stripes, bands, spots, or crossbands in a tapestry of colors. The basic growth pattern is a rosette, and most plants are about 12 inches (30 cm) across, making them fine for limited space. The flower spike bears brilliantly colored bracts from yellow to flaming red and burnt orange. The bracts last for days. The tiny flowers are white or yellow.

My *Guzmania*s grow very well at an east exposure and produce beautiful color. They are in clay pots in fir bark. I water the plants almost daily in very hot weather but do not water much in cooler weather. In any kind of weather I always keep the vases filled with water. I mist plants frequently with tepid water. Do not over fertilize these plants. The *Guzmania*s I have had reacted poorly—leaf burn can result.

Unlike most Bromeliads, *Guzmania*s do not produce abundant offshoots, producing only a few keikis, and they are generally slow growing compared to the *Billbergia*s. They include small plants, medium growers, and a few large ones to 48 inches (122 cm) across. Here is an incredible array of good indoor plants for all to try. Pests are a rarely bothersome to these Bromeliads.

41. *G. donnell-smithii*

Large plant; rosette-shaped green fan with tall red flower spike. Outstanding landscape plant. Color lasts for weeks.

42. *G. lindenii*

A 40-inch (102-cm) rosette has spectacular green foliage marked with transverse wavy lines of dark green, red beneath. The erect flower scape has green bracts and white flowers.

43. *G. lingulata*

A handsome species about 18 inches (46 cm) across with a star-shaped orange flower head.

44. *G. lingulata* var. *major*

Growing to 30 inches (76 cm) across and has a larger flower.

45. *G. lingulata* var. *minor*

Growing to 24 inches (61 cm) across and has a red-orange inflorescence with white flowers.

46. *G. musaica*

This plant is sure to please Bromeliad enthusiasts. The leaves are 24 inches (61 cm) long, bright green, banded and overlaid with irregular lines of dark green and having wavy purple markings on the reverse. The flower spike is erect and turns red at flowering time. The white waxy flowers are set tight into the poker-shaped flower head.

47. *G. sanguinea*

Native to Costa Rica and Ecuador. Brought to cultivation about 1900. A panorama of red color in leaves and crimson center at blooming time.

48. *G. scherzeriana*

A tight rosette, green-and-red leaves and a branched flower head. From Colombia and Ecuador.

49. *G. squarrosa*

A large plant, full crown of vivid red bracts. Green leaves. Flowers for two or three months. Sensitive to cool conditions.

50. *G. wittmackii*

A popular Bromeliad; greenish leaves, easy to grow, and bears a branched yellow-and-red inflorescence.

51. *G. zahnii*

Grows to 20 inches (51 cm), the delicate leaves almost transparent and striped red-brown. The flower spike has bright red bracts and white flowers that hold color for six to eight weeks.

Hechtia

These Bromeliads are perhaps more bizarre than beautiful. Plants generally have spiny leaves resembling cactus and range in size from 6 inches (15 cm) to 3 feet (91 cm). The plants coming from hot areas of Texas, Mexico, and northern Central America are terrestrial and grow in desert hillsides baking in sun, where leaves turn exquisite colors, typically pink.

In cultivation they can, if necessary, withstand cool places (to 50°F, or 10°C). Flowers are on branched spikes that come from the side of the rosette rather than from the center, as in the case of most Bromeliads.

Grow *Hechtia*s in sandy soil and allow the soil to be somewhat dry—overwatering can cause harm. Keep them in bright light or in sun if you want the bronzy red coloring that is so attractive. Repot only when necessary. These plants don't do well when roots are disturbed.

52. *H. fragilis*

Toothed leaves, green rosette. Thorny.

53. *H. glabra*

Thin green leaves, loose rosette. Wiry foliage. Good bedding plant.

54. *H. catingae* var. *horrida*

From Brazil, a handsome large plant, thorny, with branched inflorescence. Very decorative and keeps color for months. Straw-colored foliage.

55. *H. stellata*

Plants have 36- to 60-inch (91- to 152-cm) spiny golden green leaves and a tall spike with red bracts and purple flowers. An outstanding sight in any garden. Too large for most containers.

Neoregelia

In Germany and Japan, *Neoregelia*s are favorites indoors because they are handsome, undemanding, and among the finest foliage plants available. There are about sixty known *Neoregelia* species, most native to eastern Brazil; however, Colombia and Peru also have a share of these plants. In nature the plants grow near the ground or in lower branches of trees, doing best in shady places with good air circulation.

Most plants are medium sized and grow in a compact, flat rosette. Some are only a few inches across; others grow to 5 feet (152 cm). The brilliant foliage is plain green, spotted, marbled, striped, or banded with color, and at blooming time most plants show bright red to rose at the center of the plant. The flowers are small, usually in shades of blue or a combination of blue and white. Flowers die quickly but the flush of color in the foliage remains for many months. *Neoregelia*s are fine for northern exposures and excellent if you have little time to spend in care but still want some colorful plants in the home.

*Neoregelia*s will also thrive at a western exposure where they get enough light but not too much sun. I keep them at ground level so their beautiful foliage can be easily seen. I use potting mix of one third perlite, one third soil, and one third fine-grade fir bark that drains readily. I flood plants with water in the warm months, but they are given less moisture in cool weather. I rarely fertilize *Neoregelia*s, but I always keep the vase filled with water. In my plant room, small insects and frogs may visit the center of the plants—those insects that die in the plant are excellent nutrition for plants. Unfortunately, the water reservoir can attract mosquitoes, so flush vases with water weekly. Occasionally wipe leaves with a damp cloth and spray foliage with tepid water, especially in the hot weather.

*Neoregelia*s are rarely bothered by pests and are free with their offshoots, as many as four or five at a time. Plants become overcrowded in pots if you do not cut away the offsets. I purposely let this happen in a few pots to get a statement of color.

56. *N. concentrica* 'Pot Luck'

Native to Brazil, robust plant with a 28- to 30-inch (71- to 76-cm) rosette of pale green leaves marbled purple. Inner leaves turn rich red before flowering.

57. *N. carcharodon*

A Brazilian native plant with grayish green spiny leaves spotted above. Variable color.

58. *N.* 'Dexter's Pride'

Brilliant red leaves, large rosette.

59. *N. carolinae* 'Meyendorfii'

A broad 30-inch (76-cm) rosette of flat olive green leaves with coppery tinting. At flowering time the inner leaves turn a dark maroon; flowers are lilac and deep in the center.

60. *N.* 'Stars 'N Bars'

A brilliant rosette of color with large rosette. Center purple at bloom time.

61. *N. marmorata*

A species with yellow, green, and crimson leaves. Rosette grows to 30 inches (76 cm) across. Spring- or winter-flowering, it needs good light for proper leaf color. Flowers are white and deep in the cup. This one seems to thrive on neglect. It's for people who "can't grow anything." From Brazil.

62. *N.* 'Bonfire'

Grows to 20 inches (51 cm) in diameter—a beautiful rosette of reddish plum leaves. Flower crown deep in plant; tiny violet petals.

63. *N.* 'Aztec'

Heavily mottled light green and maroon splotches—exquisite foliage plant. Flowers pinkish white.

64. *N. concentrica*

About 30 inches (76 cm) across, has pale green leaves flecked with purple and edged with black spines. Leaf tips are red. Before bloom, the core of the plant turns fiery red-purple and the tiny flowers are blue.

65. *N. spectabilis*

Called the Painted Fingernail plant; grows to 30 inches (76 cm) across with spine-free, leathery green leaves tipped cerise. The small blue flowers usually appear in warm weather.

66. *N.* 'Big Bands'

Pale yellow leaves in a full rosette, red tipped.

67. *N. johannis* 'DeRolf'

A durable species; 20-inch (51-cm) green leaves with a lavender center. Although not as handsome as others in the genus, it is still worthwhile.

68. *N.* 'Passion'

Rosette pale green and red. Purple-violet leaves are outstanding in the garden setting.

Nidularium

Similar to *Neoregelia*s, *Nidularium*s make excellent houseplants if you have limited time to devote to plants. They grow readily, produce handsome leaf color, and are usually small- to medium-sized plants suitable for almost any place in the home. They grow well under more light than *Neoregelia*s but otherwise need the same conditions.

The thirty-five species of *Nidularium* are from eastern Brazil. They have flat rosettes with a fan shape and at blooming time, the center of the plant turns red. Flowers are usually hidden in the center and are inconspicuous, usually red, white, or pale blue. Leaves are glossy and finely toothed; they vary in color from green to purple and may be striped or mottled.

I grow *Nidularium*s at a southwest window where they get excellent light; thus the foliage color is always handsome. I have plants in equal parts of perlite, fir bark, and soil that drains readily. I keep the center of the plant filled with water. I do not fertilize *Nidularium*s. Plants set offshoots in abundance, and the plants require very little care once they are established, almost growing by themselves. Insects are never a problem. Grouped with *Neoregelia*s, these plants make a handsome display. There are many fine new hybrids available.

69. *N. innocentii* var. *striatum*

A colorful rosette, foliage lined with bronze and small round flower crown.

70. *Nidularium* in landscape

This figure shows the use of a number of *Nidularium*s in an outdoor setting.

Orthophytum

I have rarely seen these plants in collections, yet several are very hardy subjects that can tolerate great abuse and still provide colorful accents. Generally these are small plants, good for window display or desk accent. They require only good light to thrive.

The genus *Orthphytum* consists of about twenty species from Brazil. Most are rock-growing plants that bask in direct sun. Plants resemble succulents or cacti because of their spiny and grooved, crinkled, or striated leaves. There are several growth forms, but the rosette type predominates. At blooming time the center of the plant turns fiery red, creating a dramatic picture. White flowers are borne on short stems.

If you give *Orthophytum*s sun they grow into handsome plants, but without adequate light they cannot grow well. I use potting mix of equal parts gravel, soil, and sand that drains readily. Plants can tolerate coolness if necessary, so a few nights of 50°F (10°C) will not harm them, although an optimum temperature of, say, 65°F (18°C) is best. I water my plants sparsely all year. I never flood them but I never leave them bone dry, either. These slow-growing, undemanding plants are worth a try.

71. *O. fosterianum*

A medium-sized plant, with 12-inch (30-cm) apple-green, leathery leaves edged with spines. It bears tufted flower crowns at the leaf axils. A very different Bromeliad; the best in the group.

72. *O. navioides*

With narrow, arching 10-inch (25-cm) green leaves, it makes a handsome rosette. It is small and good for a window sill.

Pitcairnia

This is a large genus of Bromeliads but only a few are cultivated. The majority originate in Colombia, Peru, and Brazil. The plants are diversified in habit and growth shape. Some are grasslike, others have succulent-type leaves. Many grow in the ground while some are tree dwellers.

As indoor subjects some of the smaller *Pitcairnia*s do well, but large species are best grown in the subtropical garden.

73. *P. burle-marxii*

A famous plant bearing the Burle Marx name. Small, from Brazil, leaves highly colored blue-green. Can take shade.

74. *P. flammea*

Usually a terrestrial plant; variable leaf coloring. Red flower bracts.

75. *P. rubronigriflora*

Large, from Brazil; long-lasting flower spike in red and black, from which it gets its species name. Green leaves, good in landscapes.

76. *P. smithiorum*

From Peru, this clustering green Bromeliad requires little light and has a long blooming period.

77. *P. kermesiana*

Grows to 26 inches (66 cm), has pink flowers on an erect, branched stalk. Reddish green leaves.

78. *P. petropolitana*

A handsome green plant, flower crown red, most popular in the genus.
Masses of pink, green, and purple flowers.

79. *P. petropolitana* var. *petropolitana*

Resembles the preceding, but somewhat shorter flower crown.

80. *P. stenorhyncha*

Medium to large, with silvery bronze foliage. Flower spike green-bronze; open rosette.

Quesnelia

*Quesnelia*s are from eastern Brazil, and the 17 species grow in great masses in sandy soils or in the rain forests in shade. The plants are tubular with mainly green leaves that have small spines. The inflorescence has bright rose bracts and pink-and-blue flowers. Most plants are only 30 inches (76 cm) or so, making them fine for gardens. The leathery texture of the foliage makes it almost impervious to insect attack. These robust subjects can tolerate abuse and still survive in most situations.

Grow *Quesnelia*s in equal parts of soil and fir bark that drains readily. Plants require a great deal of water. Keep vases filled with water at all times. A little liquid fertilizer (10-10-5) two or three times a year is fine, but do not overdo it. Spray foliage to encourage good humidity. Plants grow well in both shade or sun, but if you want the brilliant floral spikes, sun is necessary. These Bromeliads are outstanding and flower crowns stay colorful for many weeks. Plants bear several offshoots.

81. *Q. arvensis*

One of the most striking in the genus; a 30-inch (76-cm) tubular, vase shaped plant with cross-banded foliage. The inflorescence is a vivid red cone stuffed with blue and white flowers.

82. *Q. imbricata*

Small species with small flower head. Green leaves.

83. *Q. testudo*

Similar to *Q. arvensis*; silver banded leaves, variable.

Streptocalyx (Aechmea)

Large plants in bloom with red inflorescence, *Streptocalyx* species are uncommonly beautiful. If you have a garden room or greenhouse, *Streptocalyx* can become a major accent. The genus is closely allied to *Aechmea*s and have prickly, light-green leaves in rosette growth. Native to Brazil, Ecuador, Peru, and Colombia, plants grow high in trees in hot, humid conditions.

I grow my *Streptocalyx* plants in large pots in a sunny exposure where temperatures rarely drop below 60°F (15°C). They need a great deal of water and are doused almost daily in summer.

If you are seeking a decorative plant to fill a corner of a yard, a *Streptocalyx* makes an effective display. Plants are rarely bothered by insects. Mature specimens, some grow to five feet high, are especially effective outdoors. They are costly, but certainly an exciting challenge for the advanced gardener.

Note: *Streptocalyx* has recently been considered a synonym of the genus *Aechmea*. Here these plants are noted as *S.* (*Aechmea*) *longifolia*, etc.

84. *S. (Aechmea) longifolia*

Grows to 36 inches (91 cm) and has a dense rosette of spiny, rich green leaves. The ovoid inflorescence is a whitish rose color on a short scape.

85. *S. poeppigii* (*Aechmea vallerandii*)

A 40-inch (101-cm) rosette with narrow, spiny, coppery green leaves. A handsome foliage plant that bears an erect scape crowned with a cylindrical head of densely set red or rose-purple flowers followed by pink-and-white berries, which gradually change to purple and last for months. Leaf color varies according to light exposure. An essential species for the Bromeliad enthusiast.

Tillandsia

Tillandsia offers the gardener a host of plants; this very large group of Bromeliads consists of more than 500 species. Plants are variable in shape and size; some are only 1 inch (2.54 cm) across while others are 60 inches (152 cm). Most of these epiphytes don't need soil; they grow best on tree-fern slabs.

*Tillandsia*s are the predominant Bromeliads throughout tropical and subtropical Central America, Argentina, and the West Indies. Most cling to poles, rocks, or other plants; many grow in treetops where they become large colonies. Many *Tillandsia*s produce rosettes of stiff and leathery or curly leaves, but some plants are urn shaped, bulbous in growth, or covered with grayish scales that give them a frosted look. The soft-leaved species inhabit rain forests. The many silvery types are from dry areas.

*Tillandsia*s do not have basic cup growth, as do many Bromeliads, so they depend on rain for moisture. Thus, if indoors, plants must be sprayed frequently. Moisture is taken in by the leaves rather than strictly through the roots. Usually the root systems are not extensive. Flowers are tubular and the bracts are bright rose red, green, or yellow. The predominant color of flowers is blue or purple-blue. Most species are small, up to about 12 inches (30 cm), so they are fine for the gardener with limited space.

I grow a few *Tillandsia*s in fir bark in pots but most are on wired trellises or on cork bark slabs or tree knees (roots). Plants grow best in bright light. *Tillandsia*s tolerate coolness (55°F, or 12°C) but they do best in warmth (75°F, or 23°C). Do not fertilize plants as they can grow on their own once they are established. Insects rarely attack them. Gray-leaved *Tillandsia*s are sensitive to overwatering, so grow mildly moist, never wet.

Note: Generally no specific sizes are given for most *Tillandsia*s because plants vary greatly in size, and many have recurving leaves, making it difficult to estimate exact proportions.

86. *T. brachycaulos*

Handsome, with many leaves. At blooming time the center foliage turns from green to coppery red with purple flowers. Mount this species on a slab of tree fern.

87. *T. bulbosa*

Has a bulbous base with narrow, leathery green leaves. The inflorescence is pretty, magenta and white. An oddity best grown on a piece of bark or branch.

88. *T. caput-medusae*

Resembles *T. bulbosa*. Small with vivid blue flowers.

89. *T. cyanea*

One of the most popular *Tillandsia*s. It is a regal plant with graceful, arching dark-green leaves. From the center of this medium-sized species, an erect flower stalk bears a feathery pink sword of large, purple-shaded flowers. A stunning Bromeliad that needs moisture and humidity to bloom. This is one of the more difficult ones to grow but well worth a try.

90. *T. ionantha*

A dwarf, hardly more than 2 inches (5 cm) high, but it will astound you at blooming time in spring when all the leaves turn fiery red, and tiny, pretty purple flowers appear. Grow it in sun or bright light. It needs little care.

91. *T. jalisco-monticola*

Rosette of tough silver-green hard leaves. Big rosette from Mexico with a red cone that lasts in color for weeks.

92. *T. lindenii*

Similar to *T. cyanea* with long, graceful, tapered reddish green leaves. The inflorescence is pink and blue. A real beauty but difficult to bring to bloom.

93. *T. lucida*

Mainly epiphytic from Mexico. Tall branched flower head.

94. *T. streptophylla*

A medium-sized plant with a bulbous base and grayish green, curving foliage. The inflorescence is branched, almost the same color as the leaves, with pink bracts. A mature specimen is handsome.

95. *T. stricta*

A rosette of recurving, leathery gray leaves covered with silver scales; rose bracts and blue flowers.

96. *T. vernicosa*

Small species from Bolivia. Plant forms a loose rosette. Silvery foliage.

Vriesea (Alcantarea)

Years ago one plant in this group brought *Vriesea*s recognition: the Flaming Sword, *V. splendens,* which opened the door for many of the other *Vriesea*s. In Europe, *Vriesea*s have long been grown.

About 225 species of *Vriesea* inhabit a large region from Mexico, Central America, and the West Indies to Peru, Bolivia, Paraguay, northern Argentina, and Brazil. Plants grow in rain forests but are also found at altitudes of over 6,000 feet. *Vriesea*s are mainly epiphytic, anchoring to tree branches in areas of good air circulation. They need dappled light. Some of the larger plants grow as terrestrials. Most are vase-shaped rosettes similar in appearance to *Aechmea*s and *Guzmania*s. The majority are medium to large plants, about 36–50 inches (91–127 cm) across and have smooth leaves. Foliage is plain or apple green to an almost purple-green, or is barred, striped, banded, or spotted in contrasting colors.

When inflorescences develop, the plants are veritable rainbows. The bracts are highly colored—red, green, or purple—and last for weeks. The small flowers are generally in shades of yellow.

Grow *Vriesea*s in a bright place—direct sun is not needed. I grow them in fir bark and soil; some are suspended on wooden rafts (slats of wood). I frequently water them in hot weather and always keep the vases filled with water. I spray plants in the warm months with tepid water. I do not fertilize the plants. Although *Vriesea*s can tolerate some cold, they grow best under warm conditions and seem immune to most insect pests or disease. For the gardener who can't grow anything, *Vriesea*s are the answer. Many fine hybrids are available.

97. *Mezobromelia* (*Vriesea*) *capituligera*

Small plant with red bracts. Vase shaped. Bronze-green leaves.

98. *V. fosteriana*

Medium plant with bronze-green leaves and rosette growth. Grows best in partial sun.

99. *Alcantarea* (*Vriesea*) *imperialis*

A 60-inch (152-cm) rosette, this plant has green or dark wine-red leaves crowned with a tall branched inflorescence of red and yellow. A majestic Bromeliad that requires much space.

100. *V. malzinei*

With a compact rosette of claret-colored leaves shaded green, this plant rarely grows over 16 inches (41 cm) and is fine for the windowsill. Early in spring the cylindrical flower head has yellow bracts with green margins. Very easy to grow.

101. *V.* 'Mariae'

Called the Painted Feather plant, it is a somewhat small hybrid. It holds color all winter.

102. *Alcantarea* (*Vriesea*) *regina*

A vase-shaped plant of medium size with green leaves and green-red foliage.

103. *V. simplex*

Small plant with red-yellow bracts, green leaves and a semi-rosette shape.

104. *V. splendens*

Another good *Vriesea*, it grows to about 12 inches (30 cm) and is a perfect houseplant. The green foliage is mahogany-striped, and the thrusting spring and summer inflorescence is orange-colored.

105. *V.* 'Splendide'

Large with orange bracts and blotched green leaves. Semi-rosette growth.

106. *V. gigantea*

Blue-green flower bracts. Blue-gray leaves. Rosette-shaped.

107. *V. barilletii*

Medium plant with yellow-red bracts and shiny green leaves. Vase-shaped.

Quick Reference Guide

This Quick Reference Guide will help you identify Bromeliads.

Descriptions assume average conditions. Leaf color and size may vary.

With the exception of *Tillandsia* (epiphyte), other genera are terrestrial or epiphytic.

Exposure time is as follows: full sun 6 hours; partial sun 4 hours; diffuse light (bright) 2 hours; semi-shade no sun.

Plant size corresponds to height: small to 12 inches (30 cm); medium to 36 inches (92 cm); large to 60 inches (152 cm); extra-large above 60 inches.

	Plant size	Flower	Leaf	Growth	Exposure
Aechmea					
1. *A. blanchetiana (fraudulosa)*	medium	orange	golden green	tubular	partial sun
2. *A. retusa*	medium	orange-yellow	light green, gray	tubular	partial sun
3. *A. caudata var. variegata*	medium	orange	green, ivory-striped	rosette	diffuse light
4. *A. chantinii*	large	yellow, red	green, silver-banded	vase	partial sun
5. *A. chantinii* 'Silver Ghost'	large	yellow	green	vase	partial sun
6. *A. fasciata*	medium	blue	olive green, banded	vase	diffuse light
7. *A.* 'Foster's Favorite'	large	blue	wine red	vase	diffuse light
8. *A. fulgens*	large	violet	olive green, purple	vase	diffuse light
9. *A. gamosepala*	small	pink	green	vase	diffuse light
10. *A.* 'Moorei'	medium	rose and orange	bronze-green	rosette	partial sun
11. *A. nidularioides*	medium	red	maroon	rosette	partial sun
12. *A. nudicaulis* 'Variegata'	medium	red and yellow	gray-green, silver	tubular	partial sun
13. *A. orlandiana* 'Burning Ember'	medium	white, yellow	green, brown-banded	vase	full sun
14. *A. racinae*	small	red, yellow-black	green	vase	semi-shade
15. *A. recurvata*	small	pink	dark green	tubular	diffuse light to full sun
16. *A. woronowii*	medium	pink, coral	bronze	tubular	diffuse light
Ananas					
17. *A. comosus* 'Variegatus'	large	purple	gray-green	rosette	full sun
18. *A. nanus*	small	purple	dark green	rosette	full sun
19. *A. bracteatus var. tricolor*	large	red	bronze-green	rosette	full sun
Billbergia					
20. *B. amoena*	medium	green, blue	green	tubular	diffuse light
21. *B. elegans*	large	rose, green	gray-green	tubular	partial sun
22. *B. eloiseae*	medium	pink	silver-green	tubular	partial sun

23. B. 'Fantasia'	medium	blue	green, spotted cream or pink	tubular	partial sun
24. B. leptopoda	small	green, blue	green, spotted cream	tubular	diffuse light
25. B. lietzei	small	cerise	green	tubular	diffuse light
26. B. pallidiflora	small	pink	reddish gray, silver bands	tubular	diffuse light
27. B. pyramidalis	medium	pink	golden green	tubular	diffuse light
28. B. pyramidalis 'Kyoto'	small	red	variegated green	tubular	diffuse light
29. B. 'Strawberry Marble'	small	red, pink	blotched red and yellow	tubular	diffuse light
Bromelia					
30. B. plumieri (Karatas)	large	red	pink, green	rosette	full sun
31. B. serra 'Variegata'	medium	rose	green, white, pink	rosette	full sun
Canistrum (Wittrockia)					
32. C. (Wittrockia) gigantea	large	pale pink	green, blotched with purple	open rosette	semi-shade
33. C. (Edmundoa) lindenii var. marginata	small	whitish green	yellow-green	rosette	semi-shade
Cryptanthus					
34. C. acaulis hybrids	small	white	green	rosette	diffuse light or semi-shade
35. C. 'Cascade'	small	white	bronzy leaf	open rosette	diffuse light or semi-shade
36. C. fosterianus 'Elaine'	small	white	bronze-brown	rosette	diffuse light
37. Cryptanthus in-terrarium	-	-	-	-	-
Dykia					
38. D. fosteriana	medium	orange	silver-green	rosette	partial sun
39. D. hybrids	small	bronze	thorny gray	rosette	partial sun
40. D. microcalyx	small	bronze	gray	rosette	sun

(continued)

	Plant size	Flower	Leaf	Growth	Exposure
Guzmania					
41. *G. donnell-smithii*	small	red	green	rosette	partial sun
42. *G. lindenii*	medium	green, white	green, mottled with wavy lines, red beneath	rosette	partial sun
43. *G. lingulata*	medium	orange-red white	apple green	rosette	diffuse light or semi-shade
44. *G. lingulata* var. *major*	medium	orange, white	green	rosette	partial sun
45. *G. lingulata* var. *minor*	medium	orange	green	rosette	diffuse light or semi-shade
46. *G. musaica*	medium	golden white	green, dark green, red-brown	rosette	diffuse light or semi-shade
47. *G. sanguinea*	small	red	red	open rosette	diffuse light or semi-shade
48. *G. scherzeriana*	small	red	green and red	open rosette	diffuse light or semi-shade
49. *G. squarrosa*	small	yellow, red	bronzy leaf	open rosette	diffuse light or semi-shade
50. *G. wittmackii*	small	yellow and red	green	open rosette	diffuse light or semi-shade
51. *G. zahnii*	small	white	maroon-red	rosette	diffuse light or semi-shade
Hechtia					
52. *H. fragilis*	small	bronze	green	star-shaped	full sun
53. *H. glabra*	small	bronze	green	star-shaped	full sun
Hohenbergia					
54. *H. catingae* var. *horrida*	large	variable	straw	vase	full sun
55. *H. stellata*	large	purple	golden green	vase	full sun

Neoregelia

56. N. concentrica 'Pot Luck'	small	violet	green marbled purple	rosette	diffuse sun, semi-shade
57. N. carcharodon	large	green	silvery green	rosette	diffuse sun, semi-shade
58. N. 'Dexter's Pride'	small	red	red	rosette	diffuse sun, semi-shade
59. N. carolinae 'Meyendorfii'	medium	lilac	olive, copper	rosette	diffuse sun, semi-shade
60. N. 'Stars 'N Bars'	medium	red and violet	green, spotted	rosette	diffuse sun, semi-shade
61. N. marmorata	medium	white	yellow, green, red-marbled	rosette	partial sun
62. N. 'Bonfire'	medium	blue	plum	rosette	partial sun
63. N. 'Aztec'	small	green, red	maroon, green	rosette	partial sun
64. N. concentrica	medium	blue	pale green, purple specks, red tips	rosette	partial sun
65. N. spectabilis	medium	blue	olive green	rosette	partial sun
66. N. 'Big Bands'	medium	green	yellow leaf, red-tipped	rosette	partial sun
67. N. johannis 'DeRolf'	small	lavender, blue	green	rosette	partial sun
68. N. 'Passion'	small	pink	purple-violet	rosette	partial sun

Nidularium

69. N. innocentii var. striatum	small	pink	variegated leaves	rosette	partial sun
70. Nidularium in landscape (illustration of use)	-	-	-	-	-

Orthophytum

71. O. fosterianum	medium	white	apple green	branched rosette	diffuse light
72. O. navioides	small	white	green	rosette	diffuse light

(continued)

	Plant size	Flower	Leaf	Growth	Exposure
Pitcairnia					
73. *P. burle-marxii*	small	red	blue-green	upright	partial sun
74. *P. flammea*	medium	red	green	upright	partial sun
75. *P. rubronigriflora*	medium	red and black	green	upright	partial sun
76. *P. smithiorum*	medium	orange	green	upright	partial sun
Portea					
77. *P. kermesiana*	medium	pink	reddish green	vase	diffuse light
78. *P. petropolitana*	medium	red flower crown; pink, green, and purple flowers	green	vase	diffuse light
79. *P. petropolitana var. petropolitana*	medium	coral	green	vase	diffuse light
Puya					
80. *P. stenorhyncha*	small	white	silvery bronze	fountain	full sun
Quesnelia					
81. *Q. arvensis*	large	blue and white	dark green, banded	rosette	bright sun
82. *Q. imbricata*	small	red	green	vase	diffuse light or semi-shade
83. *Q. testudo*	medium	pink	green, silver-banded	vase	bright sun
Streptocalyx (Aechmea)					
84. *S. (Aechmea) longifolia*	medium	white	dark green	rosette	partial sun
85. *S. poeppigii (Aechmea vallerandii)*	large	rose-purple	coppery green	rosette	partial sun

Tillandsia

86. T. brachycaulos	small	purple	green turning coppery red	dense rosette	partial sun
87. T. bulbosa	small	purple, white	silver-green	bottle	diffuse sun
88. T. caput-medusae	small	blue	greenish	bottle	partial sun
89. T. cyanea	medium	violet blue	dark green	rosette	diffuse light
90. T. ionantha	small	purple	silver-green turning fiery red	rosette	partial sun
91. T. jalisco-monticola	small	red and yellow	silver-green	fountain	partial sun
92. T. lindenii	medium	pink and blue	dark reddish green	rosette	diffuse light
93. T. lucida	small	pink	greenish	fountain	diffuse light
94. T. streptophylla	medium	lilac	silver-green	bottle	diffuse light
95. T. stricta	small	blue	gray-green	bottle	diffuse light
96. T. vernicosa	small	red/yellow	silvery	fountain	diffuse light

Vriesea

97. Mezobromelia (Vriesea) capituligera	small	red	bronze-green	vase	partial sun
98. V. fosteriana	medium	red	bronze-green	rosette	partial sun
99. Alcantarea (Vriesea) imperialis	extra large	bronzy green	bright green to wine red	rosette	bright sun
100. V. malzinei	small	pink	silver-green, red	vase	bright sun
101. V. 'Mariae'	small	yellow-red	green	rosette	partial sun
102. Alcantarea (Vriesea) regina	medium	yellow-pink	green-red	vase	partial sun
103. V. simplex	small	yellow-red	green	rosette	partial sun
104. V. splendens	small	orange-red	mahogany and green, striped	rosette	partial sun
105. V. 'Splendide'	small	orange-red	black and green, blotched	rosette	bright to partial sun
106. V. gigantea	large	blue-green	blue-gray	rosette	bright sun
107. V. barilletii	small	yellow-red	green	vase	semi-shade

Guzmania berteroniana

Appendix A

Note on Bromeliad Societies

Most major cities have their own chapter of a Bromeliad society. At their meetings you can find other people interested in Bromeliads and can exchange ideas about cultivation, new species, suppliers, and the like. Many cities also hold annual Bromeliad shows. To find out about membership fees and other information, check the proper listings in your telephone book or use Internet search.

For general information on Bromeliad societies write to:
The Bromeliad Society International
Membership Secretary
6901 Kellyn Lane
Vista, CA 92084-1243
www.bsi.org

Neoregelia johannis

Appendix B

Suppliers

The following plant suppliers are those from whom I have personally purchased plants, so I can vouch for their reputation in the industry and the quality of their plants. In addition, dozens of other mail-order suppliers that include Bromeliads in their catalogs can be found in many parts of the United States. Check the Yellow Pages of local telephone books or use an Internet search. The omission of these many suppliers in no way indicates that their stock is inferior—only that I have never received any plants from them and cannot make any comments. The list is in alphabetical order.

Bullis Bromeliads
12420 S.W. 248th Street
Princeton, FL 33032

DeLeon's
13745 SW 216 St.
Goulds, FL 33170

Kent's Bromeliad Nursery
P.O. Box 2166
Vista, CA 92085

Lee Moore
P.O. Box 504B
Kendall, FL 33156

Michaels Bromeliads
973 First Dirt Road
Venice, FL 34292

Oak Hill Gardens
P.O. Box 25
Route 2 Binnie Road
Dundee, IL 60118

Seaborn's Del Dios Nursery
Route 3, Box 455
Escondido, CA 92025

Tropiflora
3530 Tallevast Road
Sarasota, FL 34243

Appendix C

Classification of Popular and Available Bromeliads

Subfamily and Genera

Bromeliodeae
  Aechmea
  Ananas
  Billbergia
  Bromelia
  Canistrum
  Cryptanthus
  Hohenbergia
  Neoregelia
  Nidularium
  Orthophytum
  Portea

  Quesnalia
  Streptocalyx (*Aechmea*)
  Wittrockia (*Canistrum*)
Tillandsiodeae
  Guzmania
  Tillandsia
  Vriesea
Pitcairinoideae
  Dykia
  Hechtia
  Pitcairnia
  Puya

Portea petropolitana
var. *extensa*

Glossary

aphid: Plant lice.

banded: Marked with bars or lines of color.

bicolored: Two-colored.

bigeneric: A cross between species of different genera.

Botrytis: A fungal disease of plants.

bract: A modified leaf.

boncolor: One color.

cultivar: A clone vegetatively propagated; cultivar names are designated by single quotation marks.

cup: A water reservoir that is created by the closeness of the leaves in the center of the plant.

discolor: Of two colors, or of different colors; may be seen used as a species name.

epiphyte: An air plant; a plant that grows on trees or other plants but is not parasitic.

genera: Plural of genus.

genus: A group of related species.

habitat: The particular area or conditions in which a plant grows.

hybrid: A cross between two species or even two hybrids.

inflorescence: The actual flowers of a plant.

keiki: An offshoot or offset; small plant, often growing close to the base of the mother plant. Hawaiian for "baby."

lanceolate: Like a lance; a narrow leaf.

margin: The edge and the area immediately adjacent to it, such as a border.

mealybug: Small white cottony plant insect.

mother plant: The original or parent plant that is producing or has produced offsets.

offset: An offshoot of the main plant; sometimes termed a keiki.

pendant: Hanging; refers to growth habit.

petiole: The stalk of a leaf.

propagation: To increase or multiply by methods of reproduction; usually by cloning.

rosette: A circular arrangement of clustering leaves.

scale: A type of insect pest.

scales: Minute, flat organs on many Bromeliads that absorb water and nutrients.

scape: The lower stem of an inflorescence.

serrated: Toothed.

spike: A compact, elongated inflorescence.

stolon: A shoot that bends to the ground.

striated: Having thin lines or bands, especially those that are parallel and close together on leaves.

tank: Same as a cup.

taxonomist: An authority who classifies and names plants.

terminal: Leaves at the end of the stem.

terrestrial: Growing in soil.

tubular: Cylindrical in shape, as in a vase.

variety: A plant that differs in notable respects from the species group to which it belongs.

vase: Same as cup, tank.

vermiculite: A sterile potting medium.

×: Indicates a cross between two named parents.

xerophytic: A plant that tolerates drought.

Acknowledgments

A book is never done alone. Many people along the way have contributed to the knowledge and information about Bromeliads presented here. A special thanks and my gratitude to Robert Alonzo, who read the manuscript and corrected errors and spellings on various Bromeliads.

Also thanks to John Banta of Florida, who allowed me to roam his beautiful Bromeliad gardens and photograph plants. One cannot talk about or write about Bromeliads without mentioning the Marie Selby Botanical Gardens in Sarasota, Florida, where I photographed many plants. A special thanks to Harry E. Luther, formerly with Selby Botanical Gardens, for his great contribution to the Bromeliad family and his continued efforts to promote this wonderful group of plants. Gratitude and thanks to the many friends whose gardens I explored for Bromeliads. And another special thanks to Bullis Bromeliads in Florida for their lovely brochures, and to Tropiflora as well, for their brochures.

Finally, as always, my thanks to the Bromeliad Society International and to members of the Caloosahatchee Bromeliad Society. And to Amanda Dubin, a special thanks for the computer work.

Bromeliad Societies can be found in many states, and I urge you to join your local Society and receive the information from the magazine offered to members. As you become familiar with Bromeliads, you may find that they become a bit of an obsession.

Bibliography

Andre, E. *Bromeliaceae Andreanae.* Reprint in English from 1889 work. M. Rothenberg and R. W. Read, eds. Pacifica, Calif.: Big Bridge Press, 1990

Baensch, Ulrich, and Ursula Baensch. *Blooming Bromeliads.* Nassau, Bahamas: Tropic Beauty Publishers, 1994.

Beadle, Don A. *A Preliminary Listing of All the Known Cultivar and Grex Names for the Bromeliaceae.* Lutz, Fla.: The Bromeliad Society, Inc., 1991.

Benzing, David H. *The Biology of the Bromeliads.* Eureka, Calif.: Mad River Press, 1980.

———. *Bromeliaceae: Profile of an Adaptive Radiation.* Cambridge, UK: Cambridge University Press, 2000.

Cowan, Richard S. "Bromeliaceae of the Guayana Highland." *Memoirs of the New York Botanical Garden*, 14(3) (1967).

Duval, Leon. *Les Bromeliacees.* Librairie Agricole, del La Maison Rustique, 26 Rue Jacob, Paris, France, 1896.

Foster, Mulford B., and Racine Sarasy Foster. *Brazil, Orchid of the Tropics.* Lancaster, Pa.: Jacques Cattell Press, 1945.

Gilmartin, A. J. *The Bromeliaceae of Ecuador.* Serie Phanerogamarum Monographiae 4: 1–255 (1972).

———. *Las Bromeliacias de Honduras.* Ceiba 11(2): 1–81 (1965).

Graf, A. B., *Exotic House Plants.* East Rutherford, N.J.: Roehrs Company, 1976.

———. *Exotica*. East Rutherford, N.J.: Roehrs Company, 1976.

Kramer, Jack. *Bromeliads*. New York: Harper and Rowe, 1981.

———. *Bromeliads: The Colorful House Plants*. New York: Van Nostrand, 1965.

Leme, Elton M. C., *Nidularium: Bromeliads of the Atlantic Forest (Brazil)*. Rio de Janeiro, Brazil: GMT Editores Ltda., 2000.

———. *Canistropsis—Bromeliads of the Atlantic Forest (Brazil)*. Rio de Janeiro, Brazil: GMT Editores Ltda., 1998.

———. *Canistrum—Bromeliads of the Atlantic Forest (Brazil)*. Rio de Janeiro, Brazil: GMT Editores Ltda., 1997.

Leme, Elton M. C., and Marigo Luiz Claudio (Photographs). *Bromeliads in the Brazilian Wilderness*. Rio De Janeiro, Brazil: Marigo Comunicação Visual, 1993.

Luther, Harry E., and Edna Sieff. *An Alphabetical list of Bromeliad Binomials*. Lutz, Fla.: The Bromeliad Society, Inc., 1991.

Mee, Margaret, and Maria das Graças Lapa Wanderley. *Bromaelias*. São Paulo, Brazil: Instituto de Botânica de São Paulo, 1992.

Oliva-Esteve, Francisco. *Bromeliaceae III*. Sarasota, Fla.: Selby Botanical Gardens Press, 2002.

———. *Bromeliads*. Caracas, Venezuela: Armitano Editores, 2000.

Oliva-Esteve, Francisco, and Julian A. Steyermark. *Bromeliaceae of Venezuela*. Caracas, Venezuela: Armitano Editores, 1987.

Padilla, Victoria. *Bromeliads*. New York: Crown Publishers, 1978.

———. *The Colorful Bromeliads*. Lutz, Fla.: The Bromeliad Society, Inc., 1981.

Parkhurst, Ronald W. *The Book of Bromeliads and Hawaiian Tropical Flowers*. Makawao, Hawaii: Pacific Isle Publishing Co., 2000.

Rauh, Werner. *Bromeliads for Home, Garden and Greenhouse*. Peter Temple, ed. Poole, Dorset, UK: Blandford Press, 1979.

Richter, W. *Anzucht und Kultur Der Bromeliaceen*. Stuttgart, Germany: Eugen Ulmer Verlag, 1950.

———. *Zimmerpflanzen von Heute und Morgen: Bromeliaceen*. Melsungen, Germany: Verlag J. Neumann-Neudamm, 1962.

Seaborn, Bill. *Bromeliads*. Laguna Hills, Calif.: Gick Publishing, 1997.

Smith, Lyman B. *The Bromeliaceae of Colombia*. Robert J. Downs, Illus. Baltimore, Md.: Reed Herbarium, 1977.

———. "The Bromeliaceae of Peru." In *Flora of Peru*, J. F. Macbride, ed., part 1, no. 3. Chicago: Field Museum of Natural History, 1936.

———. *Notes on Bromeliaceae, I-XXXIII*. Compiled and indexed by Clyde F. Reed. Baltimore, Md.: Reed Herbarium, 1971.

———. "The Bromeliaceae of Brazil." In *Smithsonian Miscellaneous Collections*, 126(1), 1955. Reprinted in *Contribution from Reed Herbarium*, XXVI. Baltimore, Md.: Reed Herbarium, 1977.

———. *Studies in the Bromeliaceae*. No. I-XVII. Reprinted from 1930–1946 editions. Clyde F. Reed, ed. In *Contributions from the Reed Herbarium*, XXVIII. Baltimore, Md.: Reed Herbarium, 1977.

Smith, Lyman B., and Robert Jack Downs. *Pitcairnioideae*. (Monograph No. 14 of the Flora Neotropica series.) New York: Hafner Press, a Division of Ma cmillan, 1974.

Smith, Lyman B., and Margaret Mee (Paintings). *The Bromeliads*. South Brunswick, N.J.: A. S. Barnes & Co. (now Barnes & Noble), 1969.

Steens, Andrew. *Bromeliads for the Contemporary Gardens*. Portland, Ore.: Timber Press, 2003.

Wall, Bill. *Air Plants and Other Bromeliads*. A Wisley Handbook. Revised by Clive Innes. London, UK: Cassell Publishers for the Royal Horticultural Society, 1994.

Wilson, Robert, and Catherine Wilson. *Bromeliads in Cultivation*. Coconut Grove, Fla.: Hurricane House Publishers, 1963.

Note: Many of the older books listed are out of print; however, Internet searches may reveal copies available for purchase or for reading online. Some works may also be found in botanical reference libraries.

Index

Page numbers in *italics* refer to illustrations.

Jack Kramer is the author of numerous gardening books, and he has written for various national magazines and lectures frequently. Mr. Kramer has appeared on national television, including *TODAY, Good Morning America, The Phil Donahue Show,* and local television stations.

Other Books by Jack Kramer

Complete House Plants (2008)
100 Orchids for Florida (2006)
Women of Flowers (2005)
The Art of Flowers (2002)
A Passion for Orchids (2002)
Botanical Orchids & How to Grow Them (1998)
Orchids for the South (1994)
The World Wildlife Fund Book of Orchids (1993)